Offshore Energy Structures

Madjid Karimirad

Offshore Energy Structures

For Wind Power, Wave Energy and Hybrid
Marine Platforms

 Springer

ISBN 978-3-319-12174-1   ISBN 978-3-319-12175-8 (eBook)
DOI 10.1007/978-3-319-12175-8
Springer Cham Heidelberg New York Dordrecht London

Library of Congress Control Number: 2014954245

Printed on acid-free paper

Springer is part of Springer Science+Business Media (www.springer.com)

Preface

Due to the rapid growth of offshore renewable energy structures such as offshore wind and ocean energy devices (such as wave energy converters and tidal current turbines), the science, technology and engineering in this field are seeing a phenomenal development. However, the needed competencies and knowledge are not available in a single reference. Particularly, for hybrid marine platforms, where wind and wave energy devices are combined to use possible synergies through proper combinations, limited information is available.

Incredible progress has been made in the last decades in the advancement of offshore energy structures, especially for offshore wind applications. Now, the bottom-fixed offshore wind turbines are mature enough to compete with land-based wind turbines. This has given rise to the development of new concepts/structures for deep ocean applications. Floating wind turbines are emerging and several concepts have been commissioned, which produce electricity. Also, wave energy converters are being well-developed during the last decades and several concepts are in the stage of producing electricity. Recently, the combination of wave and wind energy devices in hybrid marine platforms has been the focus of scientists in the field of offshore technology. This generated an obvious need for a book providing the state-of-the-art knowledge in offshore energy structures.

Offshore renewable industry has planned for further booming in coming years which needs having more engineers in the "offshore energy structures" field. This is what this book is about. In this book, the author has tried to avoid sophisticated mathematical expressions. The hope is that engineers with moderate mathematical background can get a proper insight to offshore energy structures by reading this book. However, to read some parts of the book, a proper knowledge of calculus is necessary.

- The book is written for MSc students and engineers in the field of offshore technology, renewable energy, marine, ocean and coastal engineering. This book can be used in MSc-level courses in departments of civil engineering, mechanical engineering as well as ocean, coastal and marine engineering.
- This book tries to simply introduce the base for design of offshore energy structures.

- The book is about wind turbines, wave energy converters, and combined concepts such as wave-wind energy platforms.
- This is a book for masters students and engineers willing to study and work in offshore renewable energy business. In general, there is no book in the market covering all these aspects.
- The book makes a link between available standards and theoretical methods. The basics have been explained and applications in real life are exemplified. Design codes, standards and numerical tools are mentioned.
- The book is applicable for engineers working in offshore business. It is easy and simple. The author has tried to avoid complicated mathematical points while explaining the physics.
- The book covers designs applicable in industry while mentioning the practical codes and needed information.

Some of the main keywords covered in this book are listed below:

- Offshore structure
- Renewable energy
- Wave energy converters
- Wind turbines
- Floating wind turbines
- Combined wave and wind energy
- Aero-hydro-elastic
- Energy structures
- Structural dynamics
- Stochastic methods

The objectives of the book, considering the design aspects needed for offshore energy structures, are explained in the first chapter. Also, the scope of the book considering the interconnections between different chapters are highlighted in the first chapter. The book consists of the following chapters:

1. Introduction
2. Wind turbines
3. Fixed offshore wind turbines
4. Floating offshore wind turbines
5. Wave energy converters
6. Combined wave and wind power devices
7. Design aspects
8. Wave and wind theories
9. Aerodynamic and hydrodynamic loads
10. Dynamic response analyses
11. Stochastic analyses

Finally, I would like to thank my dear family (in particular, my wife) for their support which enabled me to finish this important task.

Norway   Dr. Madjid Karimirad
September 2014

Contents

Chapter 1
Introduction

1.1 Background

Offshore wind, wave, tidal, thermal gradient and salinity gradient energies are the main types of offshore and marine renewable energy resources. The renewable energies have been used to overcome the recent challenges to the environmental issues, climate changes, global warming, greenhouse gasses, pollution as well as shortage of hydrocarbon energy sources (i.e. oil and gas resource limitations).

Wave and wind power are not new for civilized human. These sources have been used from 1000 years ago as a power supply; examples are windmills and sail boats. The closest usage of wind energy to the current forms of wind turbines (WTs) are windmills for grinding, which is similar to horizontal axis WTs, and applications for transferring the groundwater to the earth's surface, which is similar to vertical axis WTs. The later one had been deployed 1000 years ago in Persia, Iran (Spera 1998).

In principle, wind turbine is a mechanical/electrical device in which the kinetic energy of air is transformed into electrical energy. Increase of oil/gas prices and air-water-soil pollution in the recent years made international community, including politicians, global planners and environmental health organizations, to aim deployment of more renewable energy. Offshore energy structures such as ocean current turbines (OCTs), wave energy converters (WECs) and WTs (both fixed and floating WTs) are booming (Multon 2012).

Wave energy conversion has become a growing field in the renewable energy sector. Over the past few decades, both scientific and industrial communities had shown a great interest to wave energy. As a result, many wave energy conversion devices have been developed to extract the hydro-kinetic and hydro-potential energy from wave motion (Kallesøe 2011). It is estimated that the (potential) worldwide wave power resource is around 2 TW (Drew et al. 2009).

In coming years, more concepts comprising OCTs, WECs and offshore WTs will appear. The possible synergy in hybrid concepts can be used meanwhile the foundation, mooring system and power cable can be shared, which reduce the costs.

© Springer International Publishing Switzerland 2014
M. Karimirad, *Offshore Energy Structures*, DOI 10.1007/978-3-319-12175-8_1

Currently, the price of energy produced from offshore WTs is higher than land-based WTs. The price of energy from onshore WTs is slightly higher than oil/gas. However, the gap becomes much smaller if the side effects of fossil fuels such as environmental pollution and global warming are counted. In future, using larger offshore WTs, array of WECs and OCTs, hybrid concepts based on wave, ocean current and wind energy and new approaches for cost-effective designs help to make offshore energy comparable with fossil fuels. This needs bringing down the costs, including the capital cost, operating and maintenance costs. All the mentioned costs are closely linked to design which highlights the importance of having an optimized design based on verified concepts. This requires more research and well-educated engineers knowing offshore energy structures, offshore design codes, related standards and regulations.

Wind power deployment has been doubled since 2008, approaching 300 GW of cumulative installed capacities. The wind energy is led by China (75 GW), the USA (60 GW) and Germany (31 GW). Wind power now provides 2.5% of global electricity demand, up to 30% in Denmark, 20% in Portugal and 18% in Spain. The policy support has been helpful in motivating this tremendous growth. Progress over the past 5 years has been good, especially in low-wind-resource sites, and reducing operating and maintenance costs. Land-based wind power costs range from US$ 60/MWh to US$ 130MWh (at most onshore sites). It is competitive where wind resources are strong and financing conditions are good, but still requires support in most countries. Offshore wind technology costs are decreasing (after a decade-long increase). However, they are still higher than land-based wind power costs (IEA 2013). Onshore wind energy is recognized as a mature form of reliable renewable energy. Wind energy can satisfy the global energy demands. Historically, the greatest barriers of onshore wind energy have been visual impact and noises. Offshore wind energy can solve some part of these issues. (EWEA 2009)

Two decades have passed since the first bottom-mounted offshore WT was installed in Europe, and many large-scale commercial projects are in operation now. On the other hand, a few floating offshore wind turbines (FOWT) have been installed as a pilot project in Norway and Portugal. Several technical questions such as floater optimization and power transmission system need to be solved for future large-scale projects.

Challenges when moving offshore are mainly:

1. Technical challenges
2. Social acceptance

To overcome technical challenges: test, analysis and optimization are necessary. Cost efficiency, standardization and industrialization are needed as well. Meanwhile, collaboration with fishery industry, marine navigation safety and environmental issues are affecting the social acceptance.

Table 1.1 2050 offshore wind and ocean energy targets. (Jeffrey and Sedgwick 2011; Airoldi et al. 2012)

Offshore wind	European target	350 GW
	International target	1150 GW
Ocean energy (i.e. wave and tidal energies)	European target	188 GW
	International target	748 GW

The EU renewable energy targets for 2020 and international policies make off-shore energy structures an important knowledge field for the future. Also, in long-term periods, the scenarios of offshore renewable energy deployment will become more and more important because of their very huge potential. The international and European targets for the development of offshore wind and ocean energies (such as wave and tidal) are listed in Table 1.1.

1.2   Objectives

As it is highlighted in the previous section, further renewable energy applications should be developed in the coming years, and this requires more specialists graduated in the field of offshore renewable energies. The objective of the current book is to respond to this need. The book covers the theoretical background for designing offshore energy structures while trying to convey the understanding of the physics. Complicated mathematical points are avoided, and the author tries to explain simply what the main issues for such a design are, see Fig. 1.1. The book is written in a way to cover the important points needed for proper designing of offshore energy structures (Fig. 1.1 shows most of the subjects covered in this book).

1.3   Scope

The scope of the book and interconnections between the chapters are highlighted in Fig. 1.2). First, an introduction to WTs in Chaps 2, 3 and 4, and WECs in Chap. 5 is given. The wave and wind energy devices are exemplified keeping in mind that the theoretical background is a key parameter completed by a good understanding of physics. In Chap. 6, examples of hybrid concepts combining different concepts and combined energy units comprising wave and wind are discussed. Chapter 7 considers the design aspects. The wave and wind theories are discussed in Chap. 8. Loads are explained in Chap. 9, and dynamic analysis is covered in Chap. 10. Finally, Chap. 11 provides information needed for preliminary stochastic analysis.

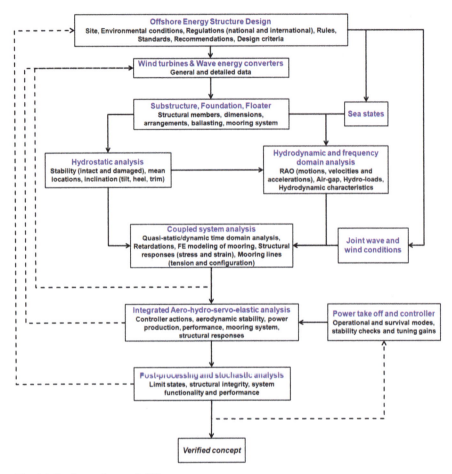

Fig. 1.1 Design analyses of offshore energy structures

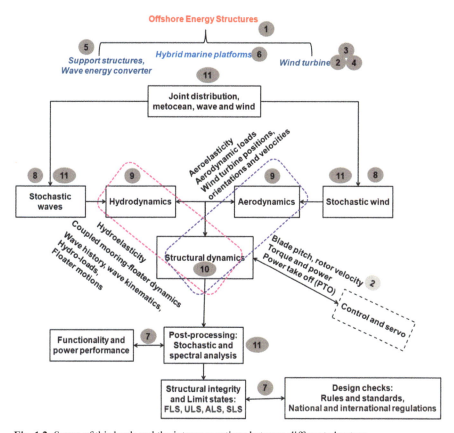

Fig. 1.2 Scope of this book and the interconnections between different chapters

References

Airoldi, D., Serri, L., & Stella, G. (2012). *Offshore renewable resources in the mediterranean area: The results of the ORECCA project*. European Seminar OWEMES 2012. Rome: OWEMES.

Drew, B., Plummer, A. R., & Sahinkaya, M. N. (2009). A review of wave energy converter technology. Proceedings of the Institution of Mechanical Engineers, Part A. *Journal of Power and Energy, 223*, 887–902.

EWEA. (2009). *Pure power-wind energy targets for 2020 and 2030*. Bruxelles, Belgium: EWEA.

IEA. (2013). *Technology roadmap*. France: International Energy Agency.

Jeffrey, H., & Sedgwick, J. (2011). ORECCA European offshore renewable energy roadmap. http://www.orecca.eu. (ORECCA coordinated action project). Accessed Aug 2014.

Kallesøe, B. S. (2011). *Aero-hydro-elastic simulation platform for wave energy systems and floating wind turbines*. Denmark: DTU, Risø-R-1767(EN).

Multon, B. (2012). *Marine renewable energy handbook*. London: Wiley.

Spera, D. A. (1998). *Wind turbine technology fundamental concepts of wind turbine*. New York: ASME.

Chapter 2
Wind Turbines

2.1 Introduction

Wind turbines generate electricity by harnessing the power of the wind. A wind turbine works the opposite of a fan (a fan uses electrical power to work). The energy in the wind turns the blades around a rotor. The rotor is connected to the main shaft, the low-speed shaft. The drive train including the gears increases the rotational speed. The high-speed shaft is connected to a generator which creates electricity. The schematic layout of a land-based wind turbine is shown in Fig. 2.1.

Based on the Rankine–Froude theory, the power (P) generated by a wind turbine can be written in the following form:

$$P = \frac{1}{2}\rho_{air} C_P A_S V_{Rel}^3 \qquad (2.1)$$

in which ρ_{air} is the air density, C_P is the power coefficient, A_S is the swept area of the wind turbine rotor and V_{Rel} is the relative wind velocity. This simple relation between the wind speed and power shows that 10 % increase in the relative wind velocity results in a 33 % increase in produced power.

2.2 Nacelle

The nacelle is located at the top of the tower (see Fig. 2.2). The nacelle is connected to the rotor and it supports several components, such as the generator and the drivetrain. For megawatt (MW) wind turbines, the nacelle is large, and some nacelles are large enough for a helicopter to land on.

The wind energy captured by the rotor is converted to electricity at the nacelle. The conversion of wind kinetic energy to electrical energy is done at the rotor nacelle. Hence, the rotor nacelle assembly is the most important part of a wind turbine.

© Springer International Publishing Switzerland 2014
M. Karimirad, *Offshore Energy Structures*, DOI 10.1007/978-3-319-12175-8_2

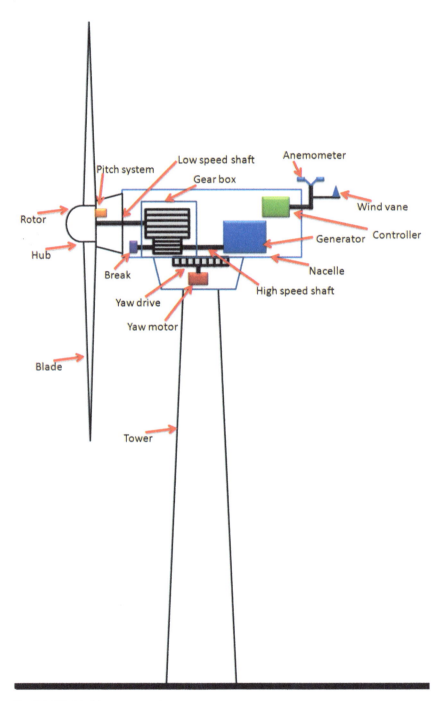

Fig. 2.1 Wind turbine components

Fig. 2.2 Nacelle of a wind turbine. (Courtesy of Paul Anderson (Anderson, Nacelle of Turbine Tower No 26—geograph.org.uk—824692.jpg, 2011b). This file is licensed under the Creative Commons Attribution-Share Alike 2.0 Generic license)

The drivetrain is a series of mechanical components, such as gears, bearings and shafts. Gearless wind turbines exist as well. However, the most developed wind turbines have gears. In a drivetrain, first is the main shaft which is connected to the rotor (hub and blades). The main shaft supports the rotor. The rear of the main shaft is connected to the slow-rotating side of the gearbox. The gearbox increases the rotational speed, e.g. 100 times.

The next component after the gearbox is the electrical generator. The generator's construction is linked to whether or not the nacelle design includes a gearbox. In front of the generator is a large disc brake that has the ability to keep the turbine in a stopped position (Lorc-website 2011).

Three-phased electrical power is generated, which must then be transformed to the higher voltage (HV) of the grid. For each phase, there is a transformer, which is placed usually at the back of the nacelle (Muljadi and Butterfield 1999).

The power cable transfers generated electricity from the generator to the grids. The cable is located in the tower. As the nacelle yaws, the cable twists to face the wind direction. The control system counts the number of cable twists to ensure that the cable is kept within defined and safe limits.

2.3 Hub

Current state-of-the-art wind turbines are using pitch-controlled variable-speed generators (E. Muljadi and C.P. Butterfield 1999). This means the pitch angle of the blades are changed to optimize the produced power. The hub is a part for the pitchable blades and their bearings, as it is clear in Fig. 2.3.

The blades are mounted on special bearings, which allow the blades to pitch, i.e. to change their angle relative to the hub while they are still in the rotor plane. The blades angle of attack (relative to the wind) can accordingly be optimized. So, the

Fig. 2.3 Hub. (Courtesy of Paul Anderson (Anderson, Hub of Turbine No 23—geograph.org.uk—837248.jpg, 2011a). This file is licensed under the Creative Commons Attribution-Share Alike 2.0 Generic license)

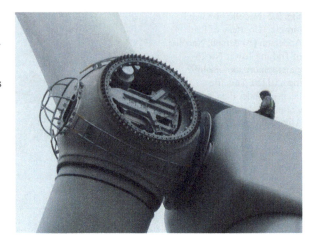

blades produce the maximum lift for a variety of wind speeds without stalling. Also, the blades are feathered to maintain the rated power when the wind velocity gets high. Fast pitching of the blades to zero degrees provides an effective means to stop the turbine (Lorc 2011).

2.4 Blades

The core of conversion of kinetic energy of wind to rotary-mechanical energy is the blade (see Fig. 2.4). A blade has an airfoil shape which directs the wind forces to the turbine low-speed shaft. Three-bladed horizontal-axis wind turbine is typical. However, downwind turbines with two blades are also used. The airfoil changes the airflow streamlines and creates pressure differences. This difference of pressure over the blade creates lift force, which is a driving force that creates torque in the wind turbine rotor. Keep in mind that drag forces appear as well, which are resistant forces and should be overcome by the structure components. The main shaft (low-speed shaft) should take these drag forces and transfer them to the nacelle base and consequently to the bottom of the tower. So, the structure should be capable of handling the drag forces together with the lift forces. The lift and drag forces are dependent on some parameters, such as the shape of the blade, the surface area, the wind speed and the angle of attack.

A symmetrical airfoil creates no lift forces when the angle of attack is zero. However, if the angle of attack is more than zero, lift occurs as a consequence of the pressure difference between the two surfaces. This discovery was made by Bernoulli and published in 1738 (Bauman 2007). Bernoulli's equation is simply a relation between static and dynamic pressure. The total pressure is assumed to be constant. Hence, when the dynamic pressure (related to the wind speed) increases, the static pressure decreases at the top of the airfoil surface. This makes a force perpendicular

Fig. 2.4 Wind turbine blades. (Courtesy of Jacopo Werther (Werther 2013). This file is licensed under the Creative Commons Attribution-Share Alike 2.0 Generic license)

to the streamlines, which is called lift. Aerodynamics will be discussed in a separate chapter of this book.

Current design solutions usually make use of three blades. However, two blades is another option. In theory, more blades over the same swept area should produce more power, but experience has shown that a design with many blades forces the wind to go around the rotor rather than through its swept area where energy can be harvested. The advantages of more than three blades are generally less than the additional costs (Lorc 2011).

2.5   Pitch System

The pitch system feathers the blades above the rated wind speed. The controller decides how much the actuators need to turn the blades. As it was mentioned earlier, it is necessary to feather the blades and control the blade angle of attack. The relative angle between the incoming wind and the blade chord should be controlled by a controller. The pitch system applies the controller's commands and feathers the blades (Jonkman et al. 2005).

The angle of attack is dependent on the wind speed, the rotational speed, and the distance from the blade root. Depending on the incoming wind velocity, there is an optimal angle of attack allowing the rotor to deliver maximum power to the main shaft. The modern wind turbine must therefore continually adjust the pitch in order to maximize energy production (Muljadi and Butterfield 1999).

Currently, the wind turbines have a collective pitch system which feathers all the blades with a same angle. Individual blade pitching can be applied in future turbines to improve the power production. Smart blades with adjustable angles of attack through the blades may be applied in a longer term. In smart-blades design, each blade is divided to several segments and each segment has its own pitching servomechanism.

Fig. 2.5 Hydraulic pitch
system. (Courtesy of Wiki-
media commons (KarleHorn
2010). This file is licensed
under the Creative Commons
Attribution-Share Alike 3.0
Unported license)

Blades of older wind turbines were directly connected to the hub with a fixed angle that could reflect the best design and power production. Those wind turbines had
no pitch control system. Hence, if the winds were strong, the blades with fixed angle
would stall as the angle of attack would increase by increasing the wind speed. This
had a positive effect on the survivability of the wind turbine and could protect the
wind turbine components, such as the blade and the tower. However, this could
reduce the amount of generated electricity. Hence, the turbine's maximum rated
production could only be reached within a narrow range of wind speeds in stall-
regulated wind turbines.

Active pitch control in modern wind turbines helps to harness energy in lower
wind speed and increases the range of wind speed in which the rated power is captured. At high wind speeds, the blades pitch to smaller angles of attack and the rated
power is maintained. This improves the survivability of the wind turbine in such
harsh conditions. Meanwhile, the power production is increased.

Depending on the turbine size, the energy needed for the servomotor to feather
the blades varies. For a 2.5-MW turbine, around 60 kW (roughly 2.5 % of the rated
power) is used. However, the gained energy from the active control of such a turbine is significant compared to the servomotor's electric consumption. Both hydraulic (Fig. 2.5) and electric systems (Fig. 2.6) can be used to feather the blades. In
offshore wind industry, the hydraulic system is the dominant option.

As it was mentioned, the blades are mounted on hub bearings; hence, they can
be feathered. Actuators mounted inside the hub can adjust the angle of attack based
on the commands coming from the central controller unit. In emergency shutdown
or fault conditions, sudden feathering of the blades can be used to stop the turbine.
This is an aerodynamic break in which the angle of attack is set rapidly to zero to
neutralize the wind forces on the blades.

Safety requirements include a pressurized tank to store sufficient energy to
stop the turbine if the central electric system fails. In some turbines, an electrical
pitch system is applied, in which the blades are feathered by gearmotors. Safety

Fig. 2.6 Electric pitch system. (Courtesy of geograph. org.uk (Anderson, Interior of the hub of Turbine No 3, 2008b). © Copyright Paul Anderson and licensed for reuse under this Creative Commons Licence)

requirements include an emergency circuit with a battery to be activated in fault conditions, i.e. when the central electric system fails.

2.6 Main Shaft

The high-speed shaft is connected to the gearbox and transmits the mechanical power of the rotor to the generator. The low-speed shaft drives the high-speed shaft through gears. The main shaft (low-speed shaft) has important functions (see Fig. 2.7). It supports the rotor (hub and blades), as well as transmits the rotary motion of the rotor and torque moments to the gearbox and/or generator. The thrust loads are taken by the shaft and transmitted to the nacelle and to the top of the tower. The low-speed shaft (main shaft) is a massive part usually built from forged or

Fig. 2.7 Main shaft. (Courtesy of geograph.org.uk (Anderson, Gearbox, Rotor Shaft and Disk Brake Assembly, 2008a). © Copyright Paul Anderson and licensed for reuse under this Creative Commons Licence)

cast iron. New materials such as carbon-fiber-reinforced polymer (CFRP) are introduced to reduce the main-shaft weight, saving several tons in the complete nacelle
(windpowermonthly.com 2012).

2.7 Gearbox

The gearbox (Fig. 2.8) connects the low-speed shaft to the high-speed shaft and increases the rotational speeds. For turbines, e.g. of 1 MW, the rotational speed of the
rotor is about 20 rotations per minute (rpm). The gearbox increases the rotational
speed by a ratio of about 90 times to get the rotational speed of the high-speed shaft
to reach about 1800 rpm. The corresponding values for a large turbine, e.g. 5 MW
(REpower 5M machine), are around 12.1, 97 and 1173.7 rpm, for rotor rotational
speed, gearbox ratio and high-speed shaft rpm, respectively.

A wind turbine may need a gear system depending on its generator type. Some
wind turbines do not need a gearbox and they are gearless (direct drivetrain). Gearbox design in a wind turbine is tightly linked to the choice of generator. Electric
generators need high rotational speed input. However, the rotor of a wind turbine
is rotating relatively slow. Hence, a gearbox system is needed to increase the rotational speed of the input torque. The low-speed shaft (main shaft) is connected to
the hub and delivers the torque to the gearbox. The gear systems, which consist of
several types of gears, increase the rotational speed roughly 100–200 times for MW
turbines (depending on the type and scale of the turbine). The high-speed shaft is
connected to the generator.

The main shaft speed is dependent on the blade tip speed and the length of the
blades. Hence, longer blades result in slower rotation of the main shaft as it is needed to keep the tip speed subsonic. The power is a multiplication of rotational speed
and torque. When a gearbox increases the rotational speed, it simultaneously decreases the torque. MW turbines have multi-pole generators, e.g. four-pole generators. In general, the less complicated is the generator; the more complicated should
be the gear system. Usually, a gear system consists of both planetary and parallel
gear stages.

Fig. 2.8 Exploded view of GRC gearbox components. (Courtesy US Department of Energy, US
Government 2014)

The gearbox is a mechanical system subjected to variable dynamic loads. The gearbox downtime is relatively high, especially for offshore units as the accessibility and availability are subjected to weather windows and environmental conditions. Hence, efforts have been made to make gearless turbines and connect the main shaft directly to the generator. However, this choice poses a considerable challenge to the generator design, significantly increasing the number of poles, complexity, size and price of this component (Lorc 2011). Hybrid concepts may be used as well including smaller gear ratios and less complex generators.

The Japanese company Mitsubishi Heavy Industries (MHI) is testing a 7-MW offshore wind turbine with a hydraulic drivetrain (mhi-global.com 2013). This avoids the need for a mechanical gearbox. Dozens of hydraulic cylinders around the main shaft compress the hydraulic fluid, which drives the hydraulic motors, and the hydraulic motors drive the generator. The advantages of such a design are expected to be considerable, ranging from less overall weight to using less expensive generators and discarding the gearbox (mhi-global.com 2013).

2.8 Generator

The generator is the main electrical part of the turbine that produces 60-Hz alternating current (AC) electricity, and it is usually an off-the-shelf induction generator. Relatively high rotational speed is required by most generators to produce electricity. There are gearless wind turbines, in which the idea is to remove the gears, and the turbines operate with "direct-drive" generators that operate at lower rotational speeds and do not need gearboxes. The gearbox is a costly (and heavy) part of the wind turbine, and removing it can have some advantages. However, more research is needed to investigate and develop a mature gearless turbine. Currently, the market is dominated by turbines using gears.

Induction generators were used in the first generation of wind turbines. Induction generators did not have any speed control that needed to match the frequency of the grid (hence, they were inexpensive). Using induction generators requires that the rotation should be speeded up. This means that the rotational speed of the turbine should be almost constant. Since the slip of the generator resulted in the speed range.

Modern wind turbines are designed for variable rotational speeds to maximize the power production and reduce the loads. This is done in a variety of ways using different generator principles and converter technologies. The introduction of a new generation of high-voltage, high-speed power electronic components allow a wide range of variable-speed operation for very-large-scale machines (Carlin et al. 2003).

In 1979, the designers set up a trade-off between a synchronous generator with a frequency converter in the stator circuit and a doubly fed asynchronous (induction) machine with a frequency converter in the rotor circuit. The doubly fed system was chosen (Carlin et al. 2003). The classic doubly fed induction generator was sufficient for the newer and larger turbines because of efficiency and relatively low converter cost. The Vestas V80/V90 2–3-MW turbines and the Repower 5–6-MW turbines use doubly fed generators (Lorc 2011).

2.9 Control

To optimize the functionality of a wind turbine, a control system is used. The controller increases the power production and limits the loads on the structural parts. Modern wind turbines apply active control to achieve the best performance. The control system consists of a number of computers which continuously monitor the condition of the wind turbine and collect statistics of operation from sensors. The controller constantly optimizes the energy production based upon a continuous measurement of mainly wind direction and speed (Muljadi and Butterfield 1999). The controller actively controls the yaw system, the blade pitch system and the generator.

The produced power is highly linked to the swept area. Hence, it is very important that the turbine rotor be straight into the coming wind. Any deviation results in reduction of power. The yaw system turns the nacelle and the rotor based on received information from the central controller. As mentioned earlier, the pitch system feathers the blades to adjust the angle of attack. The controller monitors and adjusts the angle of attack based on the wind direction and wind speed measurements.

To ensure that the produced energy is correctly sent to the electrical grid, a fast control of the generator should be performed as the electric current is alternating with a high frequency, e.g. 50 Hz. Novel control concepts have been proposed with the goal of making distribution networks more flexible by introducing active control mechanisms. Active control is expected to help with maintaining the stability of the power grid even after disturbances, loss of equipment or other unforeseen situations, by undertaking proactive actions to preserve the stability of the power network (Bouhafs and Mackay 2012) .

There are several types of sensors to measure wind characteristics such as speed and direction. The simplest one is a cup anemometer. It just measures the wind speed. It is highly fragile and does not work in cold areas as it usually ices in such harsh conditions.

Most turbines are equipped with an ultrasonic anemometer, which sends high-frequency sound waves "crossover" between the four poles, and from this it detects the phase shifting in the received signals. On the basis of this information, wind direction and velocities are calculated (Anderson et al. 2008).

LIDAR (light detection and ranging) systems are able to provide preview information of wind speed, direction and shears at various distances in front of the wind turbines. This technology provides the way for new control concepts such as feed forward control and model predictive control to increase the energy production and to reduce the loads of wind turbines. LIDAR detects coherent light reflected from the air molecules in front of the turbine and thus can predict the wind before it hits the rotor. With the LIDAR system, the wind turbine can react to changes (for example a gust) before it hits the turbine (Schlipf 2012).

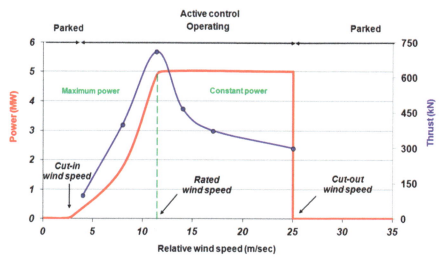

Fig. 2.9 Power and thrust curves with respect to controller phases

The controller starts up the turbine at wind speeds of about 3–5 m/s and shuts down the machine at about 25 m/s (see Fig. 2.9). Turbines do not operate at wind speeds above about 25 m/s because they may be damaged by the high winds. The power and thrust curves for a 5-MW wind turbine are shown in Fig. 2.9. The operating and parked regions are shown. The controller maintains the power above the rated wind speed to avoid excessive aerodynamic loads. Rated wind speed is a wind speed in which the wind turbine reaches its rated power, e.g. 5 MW in the current example. The cut-in wind speed is the wind speed at which the turbine starts to operate. The cut-out wind speed is the wind speed in which the controller shuts down the turbine to avoid possible damage to the structural parts due to excess of aerodynamic loads.

The brake system protects the wind turbine in emergency cases and when there is a normal shutdown. The break system acts mechanically, electrically or hydraulically depending on the design and situation.

2.10   Sensors

Monitoring is a key factor in operation of wind turbines. Sensors gather information and pass them to the controller for required actions. Some of the data are also stored for further investigation, i.e. periodic checks and maintenance. Several hundred parameter and values in the turbine are checked. Some of them are listed below:

- Rotational speed of the rotor
- Rotational speed of the generator and its voltage/current

- Lightning strikes and their charge
- Outside air temperature
- Temperature in the electronic cabinets
- Oil temperature in the gearbox
- Temperature of the generator windings
- Temperature in the gearbox bearings
- Hydraulic pressure
- Pitch angle of each rotor blade (for pitch-controlled or active-stall-controlled machines)
- Yaw angle (by counting the number of teeth on the yaw wheel)
- Number of power cable twists
- Wind direction
- Wind speed
- Size and frequency of vibrations in the nacelle and the rotor blades
- Thickness of the brake linings
- Condition of tower door, open or closed

Anemometer measures the wind speed and transmits wind speed data to the controller. The controller adjusts the generator to the torque-rotational speed of the rotor to harvest maximum possible power and feather the blades to limit the aerodynamic loads based on the wind speed.

The wind vane measures the wind direction. To produce maximum power, it is needed to the control turbine orientation with respect to the wind direction. The wind vane sends information to the yaw drive to orient the turbine.

2.11   Converter

The wind turbine converter is responsible for managing the generator. The converter controls the voltage applied by either the stator or the rotor. Usually, voltage source converters (VSC) are applied. The amount of current flowing in the generator windings and the rotating speed of the rotor are measured. Afterwards, the data are processed and the generator torque is calculated and controlled. Then, both rotational speed and electric power are regulated as electric power is current times voltage and mechanical power is torque times rotational speed.

For example: ABB's medium-voltage wind turbine converters (4–10 MW) are designed for larger turbines and provide fault ride-through and grid code compliance. They are characterized by low parts count, long life expectancy even under load cycling, high availability and low losses. The liquid-cooled converters enable low cost and efficient cable installation. ABB's low- and medium-voltage wind turbine converters are available in in-line, back-to-back or face-to-face configurations and are suitable for nacelle or tower installation (ABB 2013).

2.12 Transformer

The transformer adjusts the internal voltage of the turbine, which is typically 690 or 1000 V, to the voltage of the collector grid, which is usually 33 or 36 kV. The transformer loses power, which turns into heat and must be dissipated through the cooling system. The controller and management systems of many turbines do not take into account the transformer characteristics. The losses in the transformer have a relatively large influence on the calculated energy production. There is a tendency that the new and larger turbines use the controller to compensate the losses in the transformer.

2.13 Yaw System

The yaw drive orients upwind turbines to keep them facing the wind when the direction of wind changes. Downwind turbines do not need a yaw drive since the wind adjusts the rotor to a stabilized condition. The yaw drive is powered by a yaw motor. The yaw system is controlled by the wind turbine control system. When the wind turbine is producing electricity, these systems must continuously keep the nacelle headed directly into the incoming wind. Even a slight deviation from the correct heading will reduce the power production and increase mechanical wear on all moving parts. The control system is thus connected to a set of sensors that monitors the changing wind speed and direction at all times, making it possible to activate the yaw system quickly.

2.14 Rotor

The rotor consists of blades and a hub. It is possible to have downwind or upwind rotor configurations. Upwind turbines face into the wind while downwind turbines face away.

2.15 Tower

The tower is made from tubular steel, concrete or steel lattice. The tower supports the structure of the turbine. As wind speed increases with height, taller towers enable turbines to capture more energy and generate more electricity. The produced electrical power has a cubic relation with the velocity of the wind. For a 5-MW wind turbine, the tower height is around 90 m. The bending moments at the tower bottom are increasing by increasing the tower height. Let us consider a simplified example for illustrating the relation of tower height, structural integrity and increased weight. The static bending moment ($M_{bending}$) due to mean thrust loads can be written as:

$$M_{bending} = Thrust \times Height = 750\,kN \times 90\,m = 67.5\,MNm \qquad (2.2)$$

Then, the bending moment stress at the tower bottom can be derived as:

$$\sigma = \frac{M_{bending}R}{I_{tower}} = \frac{M_{bending}R}{\pi R^3 t} = \frac{67.5\,MNm}{\pi R^2 t} = \frac{67.5\,MNm}{\pi \times 3^2 \times 0.027} = 88\,MPa, \qquad (2.3)$$

where R is the tower section radius, I_{tower} is the area moment of inertia and t is the thickness of the tower section. To have a constant structural integrity of the base section (tower bottom) for an increased height (when shear forces and dynamics are neglected) one may consider the following:

$$\sigma = \frac{M_{bending}R}{I_{tower}} = \frac{Th \times H}{\pi R^2 t} \qquad (2.4)$$

If we neglect the increase of thrust when the height of the tower is increased, we may write a relation between tower heights (H), section thickness and the radius of the tower section.

$$\frac{H_1}{R_1^2 t_1} = \frac{H_2}{R_2^2 t_2} \qquad (2.5)$$

The weight (W) is proportional to $2\pi RtH$; hence,

$$\frac{W_2}{W_1} = \frac{R_2 t_2 H_2}{R_1 t_1 H_1}. \qquad (2.6)$$

If the thickness is kept constant:

$$\frac{W_2}{W_1} = \frac{R_2 t_2 H_2}{R_1 t_1 H_1} = \frac{R_2 H_2}{R_1 H_1} = \left(\frac{H_2}{H_1}\right)^{1.5}. \qquad (2.7)$$

If the diameter is kept constant:

$$\frac{W_2}{W_1} = \frac{R_2 t_2 H_2}{R_1 t_1 H_1} = \frac{t_2 H_2}{t_1 H_1} = \left(\frac{H_2}{H_1}\right)^2. \qquad (2.8)$$

This is extremely simplified to exemplify and show how an increase in the height of the tower can affect the weight. This gives an idea that the weight of the tower (consequently the cost) will increase with respect to the height of the tower. Detailed analyses are required to investigate optimized dimensions of a turbine for a given site.

2.16   Wind Park

Besides the wind turbine controller, each wind power plant also has a control system, the so-called wind farm controller. To get permission for connecting a wind power plant to the grid, a lot of requirements and demands to the management of the plant must be met and fulfilled. These requirements are described in the grid codes made by the transmission system operator (TSO), which is the entity of each country for transporting energy (VDN 2007).

References

ABB. (2013). ABB wind turbine converters. www.abb.com/windpower. Accessed Feb 2014.

Anderson, P. (2008a). Gearbox, Rotor Shaft and Disk Brake Assembly. http://www.geograph.org.uk/.

Anderson, P. (2008b). Interior of the hub of Turbine No 3. http://www.geograph.org.uk/photo/754033. Accessed Feb 2014.

Anderson, P. (2011a). Hub of Turbine No 23—geograph.org.uk–837248.jpg.http://commons.wikimedia.org/. Accessed Feb 2014.

Anderson, P. (2011b). Nacelle of Turbine Tower No 26–geograph.org.uk–824692.jpg. http://commons.wikimedia.org/. Accessed Feb 2014.

Anderson, D. C., Whale, J., Livingston, P. O., & Chan, D. (2008). Rooftop wind resource assessment using a three dimensional ultrasonic anemometer. http://www.ontario-sea.org/Storage/26/1798_A_Wind_Resource_Assessment_on_a_Rooftop_Using__3D_Ultrasonic_Anemometer.pdf. Accessed Feb 2014.

Bauman, R. P. (2007). The Bernoulli Conundrum. http://www.introphysics.info. Accessed Feb 2014.

Bouhafs, F., & Mackay, M. (2012). Active control and power flow routing in the Smart Grid. IEEE Smart Grid. http://smartgrid.ieee.org/december-2012/734-active-control-and-power-flow-routing-in-the-smart-grid. Accessed Feb 2014.

Carlin, P. W., Laxson, A. S., & Muljadi, E. B. (2003). The history and state of the art of variable-speed wind turbine technology. *Wind Energy, 6,* 129–159. doi:10.1002/we.77.

Jonkman, J., Marshall, L., & Buhl, M. L. Jr. (2005). *FAST user's guide.* USA: NREL/EL-500-38230.

KarleHorn. (2010). Windkraftanlage Rotorblatt Achse. de.wikipedia.

Lorc. (2011). Knowledge. http://www.lorc.dk/. Accessed Feb 2014.

Lorc-website. (2011). http://www.lorc.dk/. Accessed Feb 2014.

mhi-global.com. (2013). MHI news. http://www.mhi-global.com/discover/graph/news/no171.html. Accessed Feb 2014.

Muljadi, E., & Butterfield, C. P. (1999). *Pitch-controlled variable-speed.* Presented at the 1999 IEEE Industry Applications, Phoenix, Arizona. NREL/CP-500-27143.

Schlipf, D. (2012). Lidar Assisted Control of Wind Turbines. http://rasei.colorado.edu/wind-research-internal1/David_Schlipf.pdf. Accessed Feb 2014.

US govenment. (2014). In the OSTI Collections: Wind Power. http://www.osti.gov/home/osti_collections_wind_power.html. Accessed Feb 2014.

VDN. (2007). TransmissionCode 2007. Berlin: Verband der Netzbetreiber—VDN—e.V. beim VDEW.

Werther, J. (2013). Wind turbine blades—South Texas, USA.jpg. http://commons.wikimedia.org/.

windpowermonthly.com. (2012). Close up—Envision's 3.6 MW offshore machine. http://www.windpowermonthly.com/article/1115270/close---envisions-36mw-offshore-machine. Accessed Feb 2014.

Chapter 3
Fixed Offshore Wind Turbines

3.1 Introduction

In this chapter, a perspective of offshore wind farms, applied concepts for fixed offshore wind turbines and related statistics are given. One example of a large wind farm, which is successfully operating, is studied as well. Different concepts, their characteristics, advantages, and relative disadvantages are briefly discussed.

The European wind industry is supposed to become the most competitive energy source by 2020 onshore and by 2030 offshore. In 2009, the European Commission published "Investing in the Development of Low Carbon Technologies (SET-Plan)." The European commission stated that wind power would be capable of contributing up to 20 % of EU electricity by 2020 and as much as 33 % by 2030; this is stated in the UpWind project report (European-Commission-(FP6) 2011). The Commission's 2030 target of 33 % of the EU power from wind energy can be reached by meeting the European Wind Energy Association (EWEA)-installed capacity target of 400 GW. Based on the UpWind project report, 150 GW of this 400 GW target would be offshore wind, (European-Commission-(FP6) 2011) Currently, 9.5 GW wind power is installed offshore.

In Fig. 3.5, the turbine capacity and the number of installed offshore wind turbines based on their support structures are shown. Monopile is the most common type of almost 2000 of the installed turbine units. Jackets are supporting the maximum-rated power turbines, slightly higher than tripod wind turbines with 5 MW-rated capacity.

Floating wind turbines have recently appeared in the offshore wind market. There is no wind park based on floating structures up to now. However, some wind parks have been planned to be constructed in the near future and several scaled units are installed. The floating wind turbines are discussed in the next chapter.

© Springer International Publishing Switzerland 2014
M. Karimirad, *Offshore Energy Structures,* DOI 10.1007/978-3-319-12175-8_3

23

Fig. 3.1 Locations of the largest European operational offshore wind farms; they are mainly located in the North Sea. (google maps 2014)

3.2   Offshore Wind Farms

The locations of the largest European operational offshore wind farms are shown in Fig. 3.1. Figure 3.2 shows an example of a wind farm in the UK, Walney offshore wind farm. There are some large offshore wind parks in Asia as well. However, the largest offshore wind parks are mainly located in the North Sea. In the future, this will be expanded to the US and East Asia.

Fig. 3.2 Example of a wind farm (Walney offshore wind farm, UK). (Courtesy geo-graph.org.uk (David Dixon 2011). © Copyright David Dixon and licensed for reuse under this Creative Commons Licence)

The top largest offshore wind farms in Europe are listed in Table 3.1. The capacity of the wind farm, the country in which the farm is operating, and information about the turbine models are provided.

Also, the top largest offshore wind parks under construction are listed in Table 3.2. These numbers give an impression and feeling about the fast-booming offshore wind technology in the past years. Both Germany and the UK made a good effort in developing their capacities, and they are ahead in developing/constructing offshore wind farms, currently. One reason for this are the perfect wind resources available in those areas, especially in the North Sea.

3.3   A Case Study: Greater Gabbard Wind Farm

To have an idea about the offshore wind farms, an example is given here. Figure 3.3 shows the location of the wind farm on the map. Table 3.3 lists some information regarding this wind farm. Greater Gabbard consists of two arrays of wind turbines and associated infrastructure known as the Inner Gabbard (112 km^2) and the Galloper (35 km^2).

The timeline of the project is listed in Table 3.4 and metocean conditions (wave and wind parameters) are listed in Table 3.5. In the following chapters, the metocean, wave, and wind conditions will be explained more. The readers can check this information later and consider the case study to have a better understanding of these parameters and how they affect the design. Production and performance of the Greater Gabbard wind park are shown in Table 3.6. The capacity factor of this wind farm was 27% in 2012. The capacity factor of a wind park is the ratio of its actual output over a period of time to its potential output if it were possible for it to operate at full capacity (i.e., at rated power). The capacity factor is affected mainly by downtime of the turbines due to faults and failures plus the wind resources. For example, if there are many storms in a year, then the capacity factor will be decreased due to the shutdown of wind turbines.

$$CP = \frac{1195150 MW/(365 \times 24)}{504 MW} = 0.2707 \tag{3.1}$$

The inter array and export cables of Greater Gabbard wind farm are listed in Table 3.7. For more information refer to Greater-Gabbard-Offshore-Winds-Limited (2007) and the following websites:

http://www.lorc.dk/offshore-wind-farms-map/greater-gabbard

Greater Gabbard Offshore Wind Farm (SSE Renewables) http://sse.com/whatwedo/ourprojectsandassets/renewables/GreaterGabbard/

Greater Gabbard Offshore Wind Farm (RWE Innogy) http://www.rwe.com/web/cms/en/310132/rwe-innogy/sites/wind-offshore/in-operation/greater-gabbard/

Table 3.1 List of offshore wind farms, the top largest in Europe

Wind farm	Total (MW)	Country	Turbines & model	Official start	References
London Array (Phase I)	630	UK	175 × Siemens 3.6-120	2012	(London-array-website 2011)
Greater Gabbard	504	UK	140 × Siemens 3.6-107	2012	(offshorewind.biz, UK: Greater Gabbard offshore wind farm generates power 2012a)
Anholt	400	Denmark	111 × Siemens 3.6-120	2013	(4coffshore.com, Anholt 2013b)
BARD Offshore 1	400	Germany	80 × BARD 5.0	2013	(ndr.de 2013)
Walney (phases 1&2)	367.2	UK	102 × Siemens SWT-3.6-107	2011 (phase 1) 2012 (phase 2)	(offshorewind.biz, UK: Walney offshore wind farm fully operational 2012b)
Thorntonbank (phases 1–3)	325	Belgium	6 × REpower 5 MW, 48 × REpower 6.15 MW	2009 (phase 1) 2012 (phase 2) 2013 (phase 3)	(c-power.be, Welcome to C-Power 2013)
Sheringham Shoal	315	UK	88 × Siemens 3.6-107	2012	(Statoil-Statkraft)
Thanet	300	UK	100 × Vestas V90-3 MW	2010	(BBC-News-Technology 2010)
Lincs	270	UK	75 × 3.6 MW	2013	(smartmeters.com 2013)
Horns Rev 2	209.3	Denmark	91 × Siemens 2.3-93	2009	(dongenergy.com, About Horns Rev 2 2009)
Rodsand II	207	Denmark	90 × Siemens 2.3-93	2010	(lorc.dk 2011)
Lynn and Inner Dowsing	194	UK	54 × Siemens 3.6-107	2008	(BBC, Wind farm's first turbines active 2008)
Robin Rigg (Solway Firth)	180	UK	60 × Vestas V90-3 MW	2010	(4coffshore.com, Robin Rigg 2013d)
Gunfleet Sands	172	UK	48 × Siemens 3.6-107	2010	(dongenergy.com, About Gunfleet Sands)

Table 3.2 Top 10 under construction

Wind farm	Total (MW)	Country	Turbines and model	Completion	References
Gwynt y Môr	576	UK	160 × Siemens SWT-3.6-107	2014	(rwe.com/ 2013b)
Trianel Borkum West II	400	Germany	80 × Areva Multi-brid M5000 5 MW	2013 (I) 2015 (II)	(trianel-borkum.de 2013)
Global Tech I	400	Germany	80 × Areva multi-brid M5000 5 MW	2013	(windreich.ag 2013)
West of Duddon Sands	389	UK	108 × Siemens SWT-3.6-120	2014	(4coffshore.com, West of Duddon Sands 2013f)
Nordsee Ost	295	Germany	48 × REpower 6 M	2014	(rwe.com 2013a)
Meerwind Süd & Ost	288	Germany	80 × Siemens SWT-3.6-120	2013	(renewableenergy-focus.com 2012)
DanTysk	288	Germany	80 × Siemens SWT-3.6-120	2014	(dantysk.com 2012)
EnBW Baltic 2	288	Germany	80 × Siemens SWT-3.6-120	2014	(offshorewind.biz, Construction Starts on EnBW Baltic 2 OWF (Germany) 2013)
Amrumbank West	288	Germany	80 × Siemens SWT-3.6-120	2015	(eon.com 2013)
Borkum Riffgrund 1	277	Germany	77 × Siemens SWT-3.6-120	2015	(stateofgreen.com 2013)

Fig. 3.3 Greater Gabbard wind farm location. (google maps 2014)

Table 3.3 General information of Greater Gabbard wind farm

Operator	SSE Renewables
Installed capacity	504 MW
Number of turbines	140
Turbine	Siemens SWT-3.6-107
Development status	Commissioned
Area of wind farm	147 km^2
Layout description	The wind farm consists of two sections: inner Gabbard and Galloper sandbanks
Location	Sizewell
Region	Suffolk
Country	UK
Sea name	North Sea
Distance from shore	26 km
Water depth	24–34 m
Tidal range	0.2–4.2 m

Table 3.4 Timeline of Greater Gabbard wind farm project

Project start	2005
Construction start	2009
First power generation	2011
Commission year	2012
Developers	Greater Gabbard offshore wind:
	RWE npower renewables (50%),
	SSE renewables (50%)
Installer of turbines	A2SEA
	Seajacks
Installer of substructure	Seaway heavy lifting—Monopiles
	jumbo—transition pieces
	Red7Marine—J-tubes
Installer of inter-array cables	Subocean
	Offshore marine management (OMM)
	Technocean
	Red7Marine
Installer of export cables	Subocean
Installer of offshore substation	Seaway heavy lifting
Operator	SSE renewables
TSO	National grid
Estimated project cost	1615.25 million €

Table 3.5 Metocean, wave and wind conditions for Greater Gabbard wind farm

Average wind speed	9.0 m/s at 80 m
Average wave height	3.6 m
Significant wave height (Hs)	6.2 m

Table 3.6 Production and performance of Greater Gabbard wind farm

Total installed capacity	504 MW
Total number of turbines	140
Annual estimated production	1749 GW h/year
Annual production	1195.15 GW h in 2012
Capacity factor[a]	27.07 % in 2012

[a] The capacity factor of a wind park is the ratio of its actual output over a period of time to its potential output if it were possible for it to operate at full capacity

Table 3.7 Inter-array and export cables for Greater Gabbard wind farm

Overview	
Transmission type	MVAC/HVAC/HVAC
Operating voltage level	33 kV/132 kV/400 kV
Inter-array	
Inter-Array radials	1 radial with 18 turbines, 1 radial with 16 turbines, 1 radial with 15 turbines, 2 radials with 14 turbines, 3 radials with 11 turbines, 3 radials with 10 turbines
Inter-Array cable type	JDR cables 36 kV XLPE
Conductor size	3×240 mm^2 (Cu)
Total length	172 km
Offshore substation	
Offshore substations	2 substations
Transformers	3×180 MVA and 2×90 MVA
Electrical components supplier	Siemens transmission and distribution
Support structure type	Jackets
Foundation type	Piled
Offshore structure manufacturer	Burntisland fabrications (BiFab), one topside
	Heerema another topside
	Burtisland fabrications (BiFab), jackets
Offshore structure designer	Atkins, topside
	McNulty offshore, topside
	Rambøll, jackets

Table 3.7 (continued)

Offshore substation description	The two offshore substations are installed, respectively, in the inner Gabbard (2130 t topside) and in the Galloper (1650 t topside)
Export cable type	3 × Prysmian 132 kV XLPE
Conductor size	3 × 800 mm²
Total length	175 km
Onshore substations	
Power frequency	50 Hz
Export cable landfall	Sizewell beach, Suffolk (UK)
Onshore substation location	Sizewell, near Leiston, Suffolk (UK)

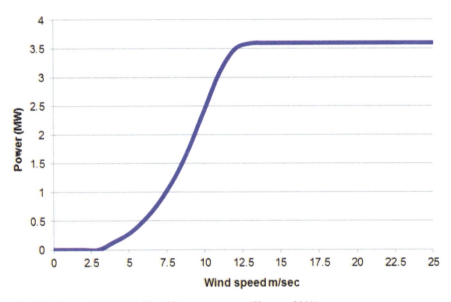

Fig. 3.4 Siemens SWT-3.6-107 turbine power curve. (Siemens 2011)

Siemens SWT-3.6-107 wind turbine is used for Greater Gabbard wind farm. The power curve of this model is shown in Fig. 3.4. The turbine specifications, such as rotor, hub, and nacelle are given in Tables 3.8 and 3.9. Tower of wind turbine and substructure (monopile) used in this wind farm are specified in Table 3.10.

Table 3.8 Turbine used in Greater Gabbard wind farm, rotor and hub

Turbine manufacturer	Siemens Wind Power
Turbine model	Siemens SWT-3.6-107
Number of turbines	140
Rated power	3.6 MW
Design life	25 years
Cut-in wind speed	4 m/s
Rated wind speed	13.5 m/s
Cut-out wind speed	25 m/s
Rotor type	3-bladed, horizontal axis
Rotor position	Upwind
Rotor diameter	107 m
Rotor area	8992 m^2
Rotor speed (minimum)	5 rpm
Rotor speed (rated)	13 rpm
Rotor weight (incl. hub)	92.5 t
Hub height (above MSL)	77.5 m
Blade tip speed (rated)	72.83 m/s
Blade tip height (above MSL)	131 m
Blade length	52 m
Blade root chord	4.2 m
Blade tip chord	1.0 m
Power regulation	Pitch regulated with variable speed (hydraulic)
Blade model	B52

Table 3.9 Turbine used in Greater Gabbard wind farm, nacelle data

Drive train type	High speed
Main bearing	Spherical roller bearing
Gearbox ratio	1:119
Gearbox stages	3 planetary stages plus 1 helical stage
Gearbox lubrication	Forced lubrication
Generator type	Asynchronous
Generator rated power	3600 kW
Generator number of poles	4 poles
Power converter location	Nacelle
Power frequency	50 Hz
Transformer voltage level	33 kV
Transformer location	Tower
Nacelle weight	142 t

Table 3.10 Structure (tower and substructure) used in Greater Gabbard wind farm

Tower	
Type	Tubular
Structure material	Steel
Height	57 m
Weight	250 t
Tower designer	Rambøll (http://www.ramboll.com/)
Substructure	
Type	Monopiles
Number of support structures	140
Support structure material	Steel
Transition piece	Weight: 300 t
Support structure	Length: 60 m and Weight: 700 t
Foundation type	Piled
Foundation structure	The monopiles are driven 30 m into seabed
Scour protection	A layer of rock, gravel or frond is used
Substructure manufacturer	Shanghai Zhenhua heavy industry (ZPMC)
Substructure designer	Rambøll

3.4   Bottom-Fixed Offshore Wind Turbine Concepts

Different concepts are proposed for offshore wind turbines. The first designs were constructed by mounting land-based wind turbine top of marine/coastal platforms. However, it was realized very soon that a proper design needs the consideration of the entire system, including foundation, support structure (platform and tower), and nacelle as one integrated unit. Now, an offshore wind turbine is not just a wind turbine mounted at the top of a marine platform. However, the first designs were based on using the knowledge of developed offshore and wind industries. In Fig. 3.5, turbine capacity and number of installed offshore wind turbines using bottom-fixed substructures are shown for different concepts. Monopiles are the most used substructures until now by a large number of installed turbines (more than 2000) around the world. Jackets are supporting the largest turbines by 5–6 MW rated power.

The terms used for defining support structures and foundation can be slightly different in literatures (see de vries et al. 2007 and DNV 2010). The main parts of an offshore wind turbine are as follows:

- Foundation and mooring system
- Support structure and platform

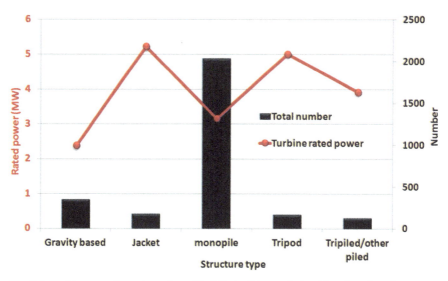

Fig. 3.5 Turbine capacity and number of installed offshore wind turbines based on the support structure type

- Transition piece
- Tower
- Nacelle (rotor, drivetrain, generator, electrical components, and housing)

In some literature, the tower is included in the support structure as it supports the rotor/nacelle assembly. In some places, the term "foundation" is used for the whole part of the installation below the tower. To be more precise, in this book, "foundation" is a part located in/on the soil and keeping the substructure in place on the seabed. The foundation has mainly three types: gravity-based, piled, and skirt/bucket (DNV 2010).

The support structure is above the foundation. The support structure includes a transition piece in some literatures. Depending on the design and concepts, the extent of transition part can be different. The tower is above the support structure, and it is connected to the transition piece. A triple wind turbine is given as an example here, see Fig. 3.6 . Based on our terminology, the following components are illustrated for this triple wind turbine:

- Pile, penetrates the soil and provides required stability
- Foundation, composed of piles in present example
- Support structure, the base for the tower and turbine
- Transition part (bolted connections), a part of the support structure connecting the tower to the base
- Tower and turbine (in general, rotor/nacelle assembly makes the turbine)

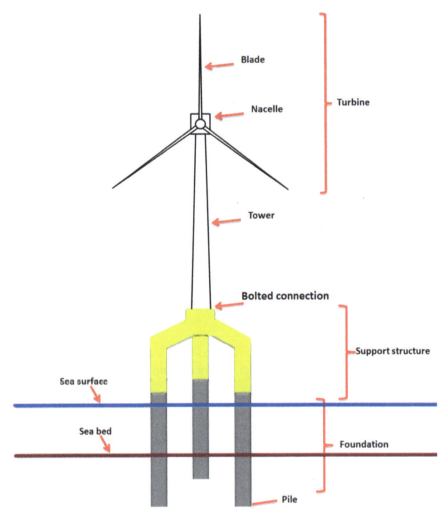

Fig. 3.6 Different parts of a tripile offshore wind turbine

3.5   Monopiles

Fixed wind turbines for offshore applications were started by using monopile in shallow water, say in 10 m water depth. A schematic layout of a monopile is shown in Fig. 3.7. The piles have been used in offshore technology for a long time, especially in jacket platforms. They are driven into the soil in order to fix the structure to the bottom of the sea. Usage of piles is also rational when it comes to offshore wind technology. For a land-based wind turbine, the tower is directly connected to the foundation, e.g., by bolting. The foundation is usually composed of steel and

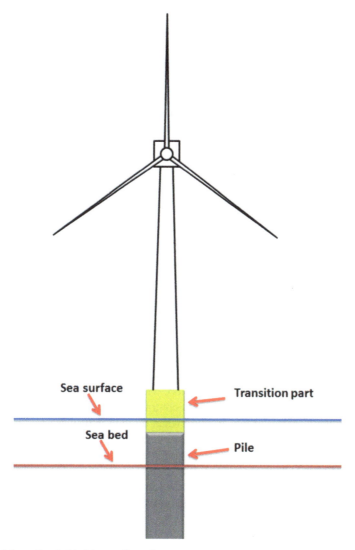

Fig. 3.7 Monopile wind turbine configuration

concrete to provide a robust base. A tower of a wind turbine for the monopile wind turbine is usually connected to the pile through a transition part.

The transition part connects the tower to the driven pile by using grout. Grouting has been applied in offshore oil-gas technology to connect the piles to the cluster of jacket platforms or in pile–leg connections. The grout can be made of seawater and cement with a water/cement weight ratio of 39%. The structural behavior of the grouted connections is rather complicated as the contact behavior of the steel grout is nonlinear. Also, the nonlinear material behavior of the grout adds to this complexity.

The grouted part is sensitive to fatigue. For a monopile, the dynamic bending moments from aerodynamic thrust can govern the grout connection due to fatigue. Finite element analyses as well as experiments can be applied to investigate the grouted connections more in detail. Such studies can be found in Honarvar et al. 2008. Monopiles are widely used for offshore wind projects in countries such as the UK, Germany, Sweden, Denmark, Belgium, and Holland. Some examples of monopile wind farm projects are Lely, Dronten/Irene Vorrink, Egmond aan Zee, Rhyl Flats, Gunfleet Sands, Belwind, Sheringham Shoal, Baltic I, Walney 1 and 2, London Array, Anholt, Thorntonbank.

Monopile wind turbines are the most common concept for offshore wind. In total, 2031 monopile wind turbines with installed a nominal capacity of 6423.9 MW are currently operating. The average rated power of each turbine is roughly 3.16 MW. A list of these turbines specifying some of their characteristics are given in Table 3.11.

3.5.1   Monopile Characteristics

More than 65 % of the offshore wind turbines have monopile supporting structure (see Table 3.11). Monopile structures are very simple in design and production. In principle, monopile is a cylinder penetrating the seabed soil. This makes the production and transportation/installation of monopiles very simple. Moreover, the analysis and engineering work needed for this type of structure is simple and well documented. All these points significantly reduce the cost. Hence, they became very popular and feasible when the offshore wind industry was born.

The simple shape of the monopile (in contrast of space frame) results in a rapid increase in the diameter of the structure when moving in deeper water to maintain the structural integrity. As the diameter increases, the structure will be subjected to more hydrodynamic loads. Hence, monopile design and feasibility of this kind of structure will be challenging in deep water, that is, above 30 m. Based on our definition in this book, a monopile is both foundation and support structure. It is the base for the wind turbine on the one hand, and on the other hand it penetrates the soil to keep the system stabilized.

The diameter of monopiles ranges from 3 to 6 m, the length of the pile is around 60 m, and almost half of the length is driven into the seabed soil. Monopiles are relatively light support structures of around 700 t (4coffshore.com, http://www.4coffshore.com/ 2013c). The thickness of the pile can be as high as 150 mm (Lorc, Knowledge 2011b; ieawind.org 2013; DNV 2010).

The transition part has usually a tubular shape. It has a slightly larger diameter than the monopile and can thus be mounted over the monopile. On top of the transition piece, a flange secures the connection with the tower using nuts and bolts. The transition piece typically weighs around 200 t and is around 25 m high (belwind.eu) and (Ballast-Nedam-Offshore 2010).

As explained above, it is easy to construct a circular, cylindrical transition piece. The axial capacity of the grouting for regular cylindrical transition piece is found to

Table 3.11 Monopile wind farms, total installed 6423.9 MW with 2031 turbines, an average of 3.16 MW for each turbine. (Lorc, Knowledge 2011b; 4coffshore. com http://www.4coffshore.com/ 2013c; ieawind.org 2013)

Name	Country	Operator	Installed Capacity	Number of Turbines	Turbine Model
Anholt	Denmark	DONG Energy	399.6 MW	111	Siemens SWT-3.6-120
Arklow Bank 1	Ireland	Airtricity	25.2 MW	7	GE 3.6 MW Offshore
Baltic 1	Germany	Energie Baden-Württemberg	48.3 MW	21	Siemens SWT-2.3-93
Barrow	UK	DONG Energy	90 MW	30	Vestas V90-3.0 MW
Belwind 1	Belgium	Belwind	165 MW	55	Vestas V90-3.0 MW
Blyth	UK	E.ON	4 MW	2	Vestas V66-2.0 MW
Bockstigen	Sweden	Nordisk Vindkraftservice	2.5 MW	5	Wind World W-3700/500 kW
Burbo Bank 1	UK	DONG Energy	90 MW	25	Siemens SWT-3.6-107
Egmond aan Zee	Netherlands	Nuon	108 MW	36	Vestas V90-3.0 MW
Greater Gabbard	UK	SSE Renewables	504 MW	140	Siemens SWT-3.6-107
Gunfleet Sands	UK	DONG Energy	172.8 MW	48	Siemens SWT-3.6-107
Gwynt y Môr	UK	RWE npower renewables	576 MW	160	Siemens SWT-3.6-107
Horns Rev 1	Denmark	Vattenfall	160 MW	80	Vestas V80-2.0 MW
Horns Rev 2	Denmark	DONG Energy	209.3 MW	91	Siemens SWT-2.3-93
Irene Vorrink	Netherlands	Nuon	16.8 MW	28	Nordtank NKT 600/43
Kamisu	Japan	Wind Power Ibaraki	14 MW	7	Subaru 80/2.0 MW
Kentish Flats 1	UK	Vattenfall	90 MW	30	Vestas V90-3.0 MW
Lely	Netherlands	Nuon	2 MW	4	Nedwind N40/500
Lincs	UK	Centrica Energy	270 MW	75	Siemens SWT-3.6-120
London Array 1	UK	DONG Energy	630 MW	175	Siemens SWT-3.6-120
Lynn and Inner Dowsing	UK	Centrica Energy	194.4 MW	54	Siemens SWT-3.6-107

Table 3.11 (continued)

Name	Country	Operator	Installed Capacity	Number of Turbines	Turbine Model
North Hoyle	UK	RWE npower renewables	60 MW	30	Vestas V80-2.0 MW
Samso	Denmark	Samso Havvind	23 MW	10	Bonus 2.3 MW/82
Prinses Amalia Windpark	Netherlands	Eneco	120 MW	60	Vestas V80-2.0 MW
Rhyl Flats	UK	RWE npower renewables	90 MW	25	Siemens SWT-3.6-107
Robin Rigg	UK	E.ON	180 MW	60	Vestas V90-3.0 MW
Scroby Sands	UK	E.ON	60 MW	30	Vestas V80-2.0 MW
Sheringham Shoal	UK	Scira Offshore Energy	316.8 MW	88	Siemens SWT-3.6-107
Teesside	UK	EDF Energy Renewables	62.1 MW	27	Siemens SWT-2.3-93
Thanet	UK	Vattenfall	300 MW	100	Vestas V90-3.0 MW
Utgrunden 1	Sweden	Vattenfall	10.5 MW	7	Enron EW 1.5s
Walney 1	UK	DONG Energy	183.6 MW	51	Siemens SWT-3.6-107
Walney 2	UK	DONG Energy	183.6 MW	51	Siemens SWT-3.6-120
Yttre Stengrund	Sweden	Vattenfall	10 MW	5	NEG Micon NM72/2000
DanTysk	Germany	Vattenfall	288 MW	80	Siemens SWT-3.6-120
Riffgat	Germany	EWE	108 MW	30	Siemens SWT-3.6-120
Meerwind Süd und Ost	Germany	WindMW	288 MW	80	Siemens SWT-3.6-120
Gunfleet Sands 3 Demonstration	UK	DONG Energy	12 MW	2	Siemens SWT-6.0-120
Northwind	Belgium		216 MW	72	Vestas V112-3.0 MW
Baltic 2	Germany	Energie Baden-Württen berg	140.4 MW	39	Siemens SWT-3.6-120

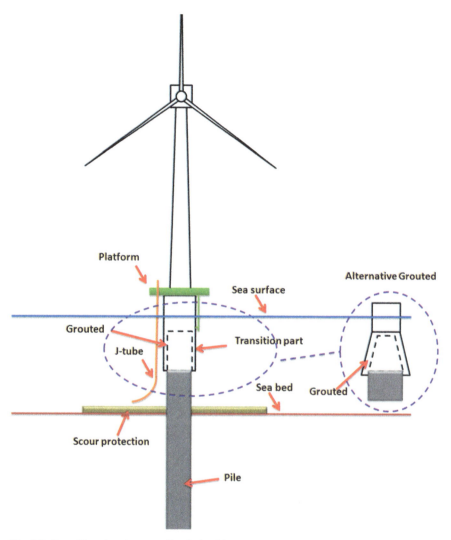

Fig. 3.8 Transition piece in monopile wind turbines

be lower than what is assumed. This is due to the effect of large diameters, the lack of control of tolerances that contributes to the axial capacity, and the abrasive wear of the grout due to the sliding of contact surfaces when subjected to large bending moments from wind and waves (dnv.com 2011). Alternative solution for transition piece is possible. For example, DNV (dnv.com 2011) has proposed a conical transition part illustrated in Fig. 3.8.

Depending on design, loads, geotechnical, and environmental conditions, the pile penetration is decided. Lateral movement, deflections, and elastic vibration affect the soil–pile interaction, as well as the grouted part and transition piece connection to the pile. Smooth transition of wave and wind loads from system to foundation needs solid and controlled filling. Otherwise, it ends with cracks and grouting crumbles. Refilling the grout/cement is necessary in some cases when the grouted part is damaged.

A limiting condition of this type of support structure is the overall deflection, lateral movement along the monopile, and vibration. Monopiles are subjected to large cyclic lateral loads and bending moments due to the current and wave loads in addition to axial loads, e.g., vertical loads due to the transition piece. Monopiles are currently the most commonly used foundation in the offshore wind market due to their ease of installation in shallow to medium water depths. This type of structure is well suited for sites with water depth ranging from 0 to 30 m.

3.6 Jacket Wind Turbine

Jacket platforms have been extensively used in offshore oil and gas applications. The oil business deployed them since the beginning of offshore oil emergence, that is, more than 50 years. They are used in different water depth with a wide range, 40–200 m. When the water depth increases, monopiles become expensive. So, the other concepts such as jacket and frame foundations (e.g. tripod) get the chance to appear in offshore wind technology when the water depth is more than 30 m. The transition depth between these concepts is not clear and highly dependent on site, resources, logistics, production cost, and owner preference. However, in most of the cases, the overall cost and net cost of the electricity are the governing parameters in decision making. Feasibility and conceptual studies can help to make such decisions. Jacket platforms are space frame structures comprising tubular elements (see Fig. 3.9).

The legs are the primary elements. In the oil/gas industry, four to eight legs are the most common. In offshore wind technology, three to four legs seem to be sufficient and practical. The legs are usually inclined. However, designs with vertical legs exist as well, e.g., with two vertical legs and two inclined legs. A transition piece connects the legs to the wind turbine tower. Concrete, steel, or hybrid can be used to make such a transition piece. The transition part is subjected to fatigue loads due to bending moments come from aerodynamic thrust and shear forces from wave loads. The design of the transition part can influence the cost of the jacket concept due to its role and weight. Hence, new research is needed to find practical solutions and proper construction methods for this component.

Braces provide stability and integrity by connecting the legs. The designs of legs are generally governed by bending moments while the designs of braces are usually governed by shear forces. In Table 3.12, the jacket wind farms are listed.

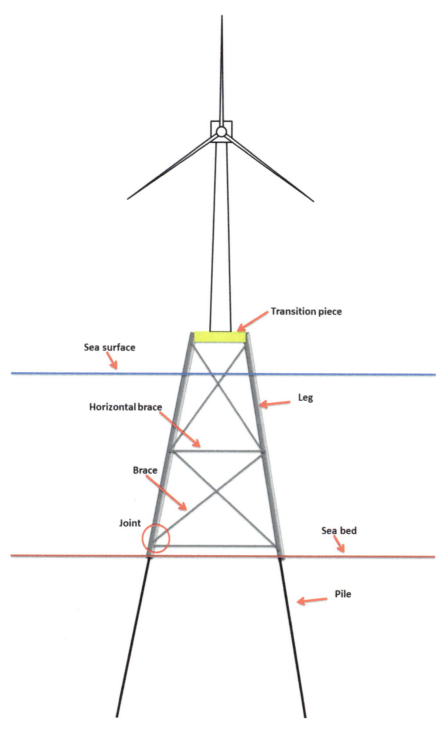

Fig. 3.9 Jacket wind turbine parts

Table 3.12 Jacket wind farms, 930.3 MW installed with 178 turbines, an average of 5.22 MW for each turbine. (Lorc, Knowledge 2011b; 4coffshore.com http://www.4coffshore.com/ 2013c; ieawind.org 2013)

Name	Country	Operator	Installed Capacity	Number of Turbines	Turbine Model
Beatrice Demonstration	UK	SSE Renewables	10 MW	2	REpower 5 M
Jeju Demonstration	South Korea	Korean Institute for Energy Research	2 MW	1	STX 72
Ormonde	UK	Vattenfall	150 MW	30	REpower 5 M
Thornton Bank 2	Belgium	C-Power	180 MW	30	REpower 6 M
Thornton Bank 3	Belgium	C-Power	108 MW	18	REpower 6 M
Suizhong Demonstration	China	China National Offshore Oil Corporation	1.5 MW	1	Goldwind GW 100/1500
Nordsee Ost	Germany	RWE Innogy	295.2 MW	48	REpower 6 M
Belwind 2 Demonstration	Belgium		6 MW	1	Alstom Haliade 150–6 MW
Alpha Ventus	Germany	Stiftung Offshore Windenergie	30 MW	6	REpower 5 M
Baltic 2	Germany	Energie Baden-Württemberg	147.6 MW	41	Siemens SWT-3.6-120

3.7 Tripile Wind Turbine

When the water depth increases, monopiles have problems to withstand wave and wind loads. The stability of the system decreases under the action of bending moments coming from wave and wind. The reason is simply due to the fact that the length of the pile increases, which results in higher bending moments. Monopiles are simple and easy to construct. This motivated to incorporate more piles to support the structure.

In tripiles, three piles are used to stabilize the system. The distance between piles provides good reaction moments and significantly helps the stability. Like monopiles, tripiles are simple structures comprising of circular cylindrical members. Their stiffness is higher than monopiles as discussed above. This makes them more proper for deeper water depth, e.g., 30–50 m. However, they are relatively heavy structures and the cost is affected by the large amount of steel and manufacturing, respectively. Table 3.13 lists the wind farms used tripile or similar piled structures for supporting the wind turbines.

The tripile structure consists of three piles (foundation), support structure, and transition piece. The piles are cylindrical steel tubes like monopiles. However, the diameter of each pile is relatively smaller compared to monopiles for the same

Table 3.13 Tripile and special piled wind farms, 505.12 MW installed with 129 turbines, each 3.91 MW in average. (Lorc, Knowledge 2011b; 4coffshore.com http://www.4coffshore.com/2013c; ieawind.org 2013)

Name	Country	Operator	Installed Capacity	Number of Turbines	Turbine Model
Rudong 1 Intertidal	China	China Longyuan Power	48.3 MW	21	Siemens SWT-2.3-101
Sakata	Japan	Summit Wind Power	16 MW	8	Vestas V80-2.0 MW
Setana	Japan	Setana Town	1.32 MW	2	Vestas V47-660 kW
Xiangshui Intertidal Demonstration	China	Yangtze New Energies Development	4.5 MW	2	SEWIND W2000M/ Goldwind GW 100/2.5 MW
Rudong Intertidal Demonstration	China	China Longyuan Power	30 MW	15	Mingyang MY 1.5s/ Guodian UP82-1500/ SEWIND W2000M/ Sany Electric SE9320III-S3/ Envision E82/CSIC HZ93-2000/BaoNan BN82-2 MW/Sinovel SL3000/90/Goldwind GW 100/2.5 MW
Bard Offshore 1	Germany	Bard	400 MW	80	Bard 5.0
Hooksiel Demonstration	Germany	Bard	5 MW	1	Bard 5.0

depth. The diameter of each pile is approximately 3 m (4Coffshore.com 2013e). Each pile can, depending on water depth and soil conditions, be up to 90 m high and weigh up to 400 t. Between 30 and 45 m of the pile rests in the soil depending on the soil properties. Three piles are connected to support the structure above the water. Each pile should be driven into the soil, separately. The support structure has three legs, which are connected to piles. At the top of the support structure, the tower is bolted. A flange-bolted connection is practical for such a purpose. The connection of the piles and the support structure legs is performed using grout/cement. The connection of the piles and the support structure legs is performed using grout/cement, e.g. grout-filled annulus of 5 m in height and 13 cm in thickness (LORC 2011c). The structure is fitted with a work platform and stairs, and the boat landing is mounted on one of the piles.

The project area for the North Sea wind farm "BARD Offshore 1" covers around 60 km^2 and is located some 90 km northwest of the island of Borkum. The water depth here is around 40 m. The offshore technicians began installing a total of 80 wind turbines of the type "BARD 5.0" in March 2010, the first of which were connected to the grid in late 2011. From September 2013, the wind turbine generators

Fig. 3.10 The Bard 5.0-MW turbines supported by tripile structures at 40-m water depth at the offshore plant "Bard Offshore 1." (commons.wikimedia.org 2013. This file is licensed under the Creative Commons Attribution-Share Alike 3.0 Unported license)

are producing a nominal capacity of 400 MW, equivalent to the electricity requirements of more than 400,000 households (bard-offshore.de 2011; see Fig. 3.10).

Some literatures called all the structural parts above piles transition piece. This terminology was being used when the offshore wind industry was born; in the beginning, the idea was to put a wind turbine over a foundation using a transition piece. New design and recent efforts consider the structural integrity of the entire system in the analyses. Hence, the transition part is getting smaller both in mind and practice.

Table 3.14 Tripod wind farms, 830 MW with 166 installed turbines, 5 MW in average for each turbine

Name	Country	Operator	Installed capacity	Number of turbines	Turbine model
Borkum West 2	Germany	Trianel	400 MW	80	AREVA M5000-116
Global Tech 1	Germany	Global Tech 1 Offshore Wind	400 MW	80	AREVA M5000-116
Alpha Ventus	Germany	Stiftung Offshore Windenergie	30 MW	6	AREVA M5000-116

3.8 Tripod Wind Turbine

The tripod structure is a relatively lightweight three-legged steel jacket compared to a standard lattice structure. These space frame structures have a steel central column, which is below the turbine. The loads from the turbine are transferred to the steel frame and consequently to piles. Piles (typically with diameter of 0.9–1.0 m) are installed at each leg position to anchor the tripod to the seabed. The three piles are driven 10–20 m into the seabed. The tripod can also be installed using suction buckets. Suction buckets are acting as the foundation instead of piles (4Coffshore. com 2013c). A list of current offshore wind farms based on a tripod is presented in Table 3.14.

The tripod foundation has good stability and overall stiffness. Like tripiles, tripods use the advantage of footprint distance to increase the stability. Hence, the water depth can increase up to 50 m. Tripod support-structure weight (without piles) is approximately 700 t (alpha-ventus.de 2010). As it is explained for tripiles, the resistance is increased due to the increased arm (footprint) between the piles. Tripods have complex main joints that increase the risk of failure due to fatigue. The tripod is a space frame structure resembling a simple lattice structure. This makes them suitable for deeper water as the stiffness and stability are increased with the help of separated piles. The foundation is cheaper than complex jackets. Also, the scour is less significant compared to monopiles. Uneven seabed can be a challenge for tripods. As it is discussed earlier, the tripod has large main joints which require precise study of fatigue life.

One of the main differences between tripods and tripiles are the wave loads. The piles in a tripile wind turbine are extending above the sea level. However, tripods have a central column which is connected to braces and through them to piles. This makes the diameter of the central column in tripods large and hence increases the wave loads. In Fig. 3.11, the layout of a tripod wind turbine is illustrated. It is not recommended to install tripods in very shallow water, e.g., 8 m; this causes problems to the vessels approaching the foundation as sufficient draught is needed to clear the steel frame(4Coffshore.com 2013e).

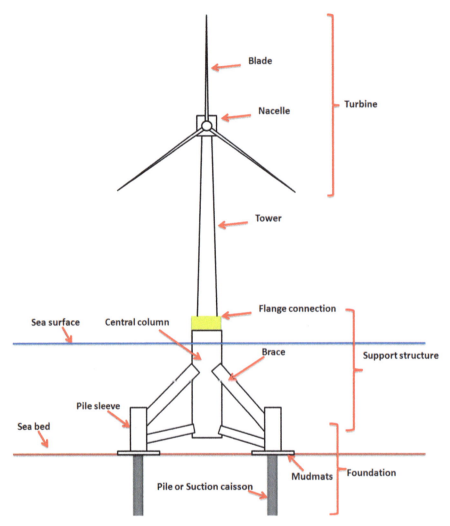

Fig. 3.11 Tripod wind turbine layout

3.9   Gravity-Based Wind Turbines

It can be surprising to know that the first offshore wind turbines were built using the gravity-based structure. Figure 3.12 shows the layout of a gravity-based wind turbine. The first offshore wind farm pilot project in the world contained 11 large concrete structures weighing in average 908 t. The structures were placed in the water near the shore of Lolland, Denmark back in the year 1991. Vindeby wind farm was the largest offshore wind farm until 2000 when Middelgrunden was constructed (seas-nve.dk; see Table 3.15). A list of gravity-based offshore wind farms are presented in Table 3.16.

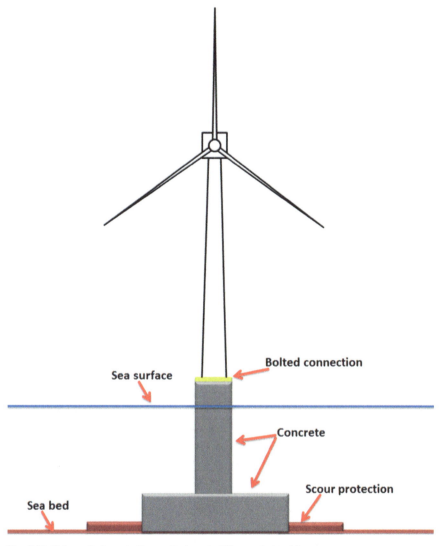

Fig. 3.12 Schematic layout of a gravity-based offshore wind turbine

Gravity-based structures are stabilized by the weight of the structure and ballast. As it is clear in the schematic layout of the structure, the bottom of the support structure holds a large amount of weight. The large area of the bottom, which is spread over the seabed helps to provide effective bending moment resistant. The advantage of this type of structure is that no drilling or hammering into the soil is needed. However, the seabed has to be prepared with dredging, gravel, and concrete. This means that they can be used regardless of seabed properties as the seabed should be prepared anyway. They usually do not require any transition piece, and the tower can be bolted at the top of the cylindrical part coming out of the water. Concrete is

Table 3.15 Key figures of Vindeby offshore wind farm. (seas-nve.dk)

Number of turbines	11
Manufacturer	Bonus energy A/S
Location	Off the north coast of the Danish Island Lolland
Annual production	11,200 MWh
Generator output per turbine	450 kW
Height of hub above the sea	35 m
Rotor diameter	35 m
Construction period	1990–1991
Construction budget in EUR	10 million

Table 3.16 Gravity-based wind farms, 836 MW with 350 installed turbines, average 2.38 MW for each turbine. (Lorc, Knowledge 2011b; 4coffshore.com http://www.4coffshore.com/2013c;. ieawind.org 2013)

Name	Country	Operator	Installed Capacity	Number of Turbines	Turbine Model
Avedore Holme	Denmark	DONG Energy	10.8 MW	3	Siemens SWT-3.6-120
Breitling Demonstration	Germany	WIND-projekt	2.5 MW	1	Nordex N90/2500 (Offshore)
Donghai Bridge 1	China	Shanghai Donghai Wind Power	102 MW	34	Sinovel SL3000/90
Ems Emden	Germany	ENOVA	4.5 MW	1	Enercon E-112
Kemi Ajos	Finland	Innopower	30 MW	10	WinWind WWD-3
Lillgrund	Sweden	Vattenfall	110.4 MW	48	Siemens SWT-2.3-93
Middelgrunden	Denmark	DONG Energy	40 MW	20	Bonus 2.0 MW/76
Nysted 1	Denmark	DONG Energy	165.6 MW	72	Bonus 2.3 MW/82
Pori Offshore 1	Finland	Suomen Hyötytuuli	2.3 MW	1	Siemens SWT-2.3-101
Rodsand 2	Denmark	E.ON	207 MW	90	Siemens SWT-2.3-93
Sprogo	Denmark	Sund & Bælt	21 MW	7	Vestas V90-3.0 MW
Thornton Bank 1	Belgium	C-Power	30 MW	6	REpower 5M
Tuno Knob	Denmark	DONG Energy	5 MW	10	Vestas V39-500 kW
Vindeby	Denmark	DONG Energy	4.95 MW	11	Bonus 450 kW/37
Vindpark Vanern	Sweden	Vindpark Vänern	30 MW	10	WinWind WWD-3
Kitakyushu Demonstration	Japan	Electric Power Development Co (J Power)	2.0 MW	1	JSW J82-2.0

Table 3.16 (continued)

Name	Country	Operator	Installed Capacity	Number of Turbines	Turbine Model
Choshi Offshore Demonstration Project	Japan	Toyko Electric Power Company	2.4 MW	1	MWT92 2.4 MW
Kårehamn	Sweden		48 MW	16	Vestas V112-3.0 MW
Ronland	Denmark	Vindenergi/ Harboøre Møl-lelaug + Thy-borøn-Harboøre Vindmøllelaug	17.2 MW	8	Vestas V80-2.0 MW/ Bonus 2.3 MW/82

Fig. 3.13 Troll A: Norwegian oil platform. (Ranveig 2005. This file is licensed under the Creative Commons Attribution-Share Alike 3.0 Unported license)

a durable material used in offshore oil and gas platforms as well. Troll A platform is an example of such designs operated by Statoil in a Norwegian field close to Bergen (see Fig. 3.13). The Troll A platform has an overall height of 472 m, weighs 683,600 t and has the distinction of being the tallest structure ever moved by mankind.

One of the other advantages of gravity-based structures is that they can be transported to the offshore site afloat. Beside the advantages, there are some disadvantages as well for gravity-based structures, e.g., they are relatively heavy structures and also become more expensive to install when the water depth increases. Usually, gravity-based structures are made without steel reinforcement. When the water depth increases, this type of structure is cost-wise not competitive with other types.

Despite the high cost of gravity-based structures in deep waters, the Thornton Bank field in Belgium, constructed in a site with water depth of 27.5 m. The support structure is roughly 42 m high and weighs around 3000 t (c-power.be (n. d.), construction of the gravity-based foundation). Sand is filled inside the structure as ballast.

References

4coffshore.com. (2013b). Anholt. http://www.4coffshore.com/windfarms/anholt-denmark-dk13.html. Accessed April 2014.
4coffshore.com. (2013c). http://www.4coffshore.com/. http://www.4coffshore.com/. Accessed April 2014.
4coffshore.com. (2013d). Robin Rigg. http://www.4coffshore.com/windfarms/robin-rigg-united-kingdom-uk20.html. Accessed April 2014.
4Coffshore.com. (2013e). Tripile support structures. http://www.4coffshore.com/windfarms/tripile-support-structures-aid272.html. Accessed April 2014.
4coffshore.com, (2013f), West of Duddon Sands. http://www.4coffshore.com/windfarms/west-of-duddon-sands-united-kingdom-uk33.html. Accessed April 2014.
alpha-ventus.de. (2010). Alpha ventus—the first German offshore wind farm. http://www.alpha-ventus.de/index.php?id=120. Accessed April 2014.
Ballast-Nedam-Offshore. (2010). Baltic 1 offshore wind farm. http://www.bnoffshore.com/public/offshore/Documents/Baltic%201%20Offshore%20Wind%20Farm.pdf. Accessed April 2014.
bard-offshore.de. (2011). BARD offshore 1. http://www.bard-offshore.de/en/projects/offshore/bard-offshore-1.html. Accessed April 2014.
BBC. (2008). Wind farm's first turbines active. http://news.bbc.co.uk/2/hi/uk_news/england/lincolnshire/7388949.stm. Accessed April 2014.
BBC. (2010). New UK offshore wind farm licences are announced. http://news.bbc.co.uk/2/hi/business/8448203.stm. Accessed April 2014.
commons.wikimedia.org. (2013). Bard offshore. http://commons.wikimedia.org/wiki/File:Bard_Offshore_2011.JPG. Accessed April 2014.
c-power.be. (2013). Welcome to C-power. http://www.c-power.be/English/welcome/algemene_info.html. Accessed April 2014.
c-power.be. (n.d.). Construction of the gravity base foundation. http://www.c-power.be/construction. Accessed April 2014.
dantysk.com. (2012). DanTysk offshore wind. http://www.dantysk.com/wind-farm/facts-chronology.html. Accessed April 2014.
de vries, W. E., et al. (2007). Assessment of bottom-mounted support structure types. UpWind Project.
Dixon, D. (2011). Walney offshore windfarm. http://www.geograph.org.uk/photo/2391702. Accessed April 2014.
DNV. (2010). Design of offshore wind turbine structures, DNV-OS-J101. Det Norske Veritas.

dnv.com. (2011). New design practices for offshore wind turbine structures. http://www.dnv.com/press_area/press_releases/2011/new_design_practices_offshore_wind_turbine_structures.asp. Accessed April 2014.

dongenergy.com. (2009). About horns rev 2. http://www.dongenergy.com/. Accessed April 2014.

eon.com. (2013). E.ON lays groundwork for Amrumbank West. http://www.eon.com/en/media/news/press-releases/2013/5/29/eon-lays-groundwork-for-amrumbank-west.html. Accessed April 2014.

European-Commission-(FP6). (2011). UpWind, design limits and solutions for very large wind turbines. The sixth framework programme for research and development of the European Commission (FP6).

googl maps. (2014). https://maps.google.com/maps?q= http://tools.wmflabs.org/kmlexport/%3Farticle%3DList_of_offshore_wind_farms%26usecache%3D1. http://en.wikipedia.org/wiki/List_of_offshore_wind_farms. Accessed April 2014.

google maps. (2014). https://www.google.com/maps/@51.879865,1.94,7z. googel.com. Accessed April 2014.

Greater-Gabbard-Offshore-Winds-Limited. (2007). Decommissioning programme, Greater Gabbard offshore wind farm project. 577000/403– MGT100– GGR – 107.

Honarvar, M. R., Bahaari, M. R., & Asgarian, B.. (2008). Experimental modeling of pile-leg interaction in jacket type offshore platforms cyclic inelastic behavior. *American Journal of Applied Sciences, 5*(11), 1448–1460.

ieawind.org. (2013). IEA Wind. http://www.ieawind.org/. Accessed April 2014.

London-array-website. (2011). First foundation installed at London Array. http://www.londonarray.com/. Accessed April 2014.

LORC. (2011c). Tripile—Three monopiles in one. http://www.lorc.dk/offshore-wind/foundations/tripiles. Accessed April 2014.

lorc.dk. (2011). Roedsand-2. http://www.lorc.dk/offshore-wind-farms-map/roedsand-2. Accessed April 2014.

ndr.de. (2013). Rösler eröffnet Offshore-Windpark Bard 1. http://www.ndr.de/regional/niedersachsen/oldenburg/offshore389.html. Accessed April 2014.

offshorewind.biz. (2012a). UK: Greater Gabbard offshore wind farm generates power. http://www.offshorewind.biz/2012/09/07/uk-greater-gabbard-offshore-wind-farm-generates-power/. Accessed April 2014.

offshorewind.biz. (2012b). UK: Walney offshore wind farm fully operational. http://www.offshorewind.biz/2012/06/14/uk-walney-offshore-wind-farm-fully-operational/. Accessed April 2014.

offshorewind.biz. (2013). Construction Starts on EnBW Baltic 2 OWF (Germany). http://www.offshorewind.biz/2013/08/20/construction-starts-on-enbw-baltic-2-owf-germany/. Accessed April 2014.

Ranveig. (2005). Oil platform Norway. http://commons.wikimedia.org/wiki/File:Oil_platform_Norway.jpg. Accessed April 2014.

renewableenergyfocus.com. (2012). Construction starts on Germany's €1.3bn Meerwind offshore wind farm project. http://www.renewableenergyfocus.com/view/27468/construction-starts-on-germany-s-1-3bn-meerwind-offshore-wind-farm-project/. Accessed April 2014.

rwe.com. (2013a). Offshore wind farm Nordsee Ost. http://www.rwe.com/web/cms/en/961656/offshore-wind-farm-nordsee-ost/. Accessed April 2014.

rwe.com/. (2013b). Gwynt y Môr website. http://www.rwe.com/web/cms/en/1202906/rwe-innogy/sites/wind-offshore/under-construction/gwynt-y-mr/. Accessed April 2014.

Siemens (2011). New dimensions, Siemens Wind Turbine SWT-3.6-107. Published by and copyright © 2011:Siemens AG Energy Sector Freyeslebenstrasse 1 91058 Erlangen, Germany.

smartmeters.com. (2013). Lincs wind farm now fully operational. http://www.smartmeters.com/the-news/renewable-energy-news/4169-lincs-wind-farm-now-fully-operational.html. Accessed April 2014.

stateofgreen.com. (2013). DONG energy reaches milestone at Borkum Riffgrund 1. http://www.
    stateofgreen.com/en/Newsroom/DONG-Energy-Reaches-Milestone-at-Borkum-Riffgrund-1.
    Accessed April 2014.
trianel-borkum.de. (2013). Windpark. http://www.trianel-borkum.de/de/windpark/daten-und-
    fakten.html. Accessed April 2014.
windreich.ag. (2013). Global Tech I—Construction Progress. http://windreich.ag/en/global-tech-i-
    construction-progress/. Accessed April 2014.

Chapter 4
Floating Offshore Wind Turbines

4.1 Introduction

In this chapter, a perspective of floating offshore wind technology, applied concepts, and related statistics are given. Floating wind turbines are recently appeared in the offshore wind market. There is no wind park based on floating structures until now. However, some wind parks are planned to be constructed in near future. Several scaled units are installed. Some examples of commissioned floating wind turbines are discussed in this chapter.

When the water depth increases, the cost associated with bottom-fixed concepts increases rapidly. Some concepts like monopile and gravity-based structures are more affected by depth increase. The other concepts like jackets came to picture to answer the need of harvesting energy in deeper water. However, in practice, the cost of application of them will not guaranty the low cost of produced electricity. Hence, offshore wind technology started to explore the feasibility of application of floating wind turbine units in deep waters, e.g., 150 m. Figure 4.1 illustrates the rated-power relation with water depth for different concepts, floating versus fixed wind turbines.

4.2 Floating Offshore Wind Projects

Among the support structures, floating structures are less used. However, they will become more popular as the industry explores offshore sites with larger water depth. When the water depth increases, the cost of using bottom-fixed turbines increases rapidly. There are several debates and researches around the world to identify the transition depth when floating platforms are economical with respect to the bottom-fixed turbines. The transition depth is in the order of 50–100 m. Depending on the type of floater and site specification, the transition depth can be influenced, and a floating concept can be feasible or not for a defined depth. However, it is clear that when the water depth is more than 100 m, the floating concepts are likely the most cost-effective solutions.

© Springer International Publishing Switzerland 2014
M. Karimirad, *Offshore Energy Structures,* DOI 10.1007/978-3-319-12175-8_4

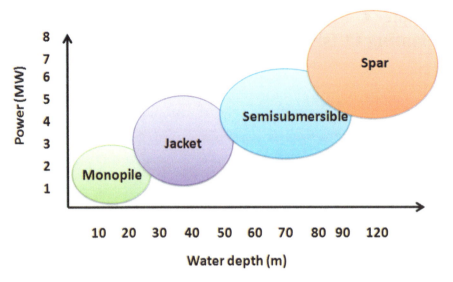

Fig. 4.1 Rated power versus water depth for different concepts, floating and fixed wind turbine concepts

There are several types of floating wind turbines inspired from offshore oil and gas industry. Most of these concepts are in the feasibility study phase or scaled wind turbine test. Model tests in ocean basins and hydrodynamic laboratories have been conducted for some concepts. Numerical simulations and benchmarking to study the concepts and developing proper analytical codes were extensively performed during the past years.

Different joint research projects had been carried out by participating scientists and researchers around the world, among them, European, Asian, and American partners were deeply involved. All these points highlight the importance of floating structures in future offshore wind business.

Currently, around 32 floating offshore projects are active globally. Semisubmersible is the most popular at the moment, and more than 13 projects are applying semisubmersible as the base floater (Fig. 4.2).

In Table 4.1, three floating offshore wind turbines installed with turbine-rated power higher than 1 MW are listed. In the coming sections, these floating offshore wind projects are discussed in detail. There are several research concepts and small-scaled floating turbines installed, which will be discussed later.

4.3   Hywind Project

Hywind is the world's first full-scale floating wind turbine, see Fig. 4.3. In 2009, Statoil accomplished launching of new technology: floating wind technology, see Table 4.2 for more information. Several attempts for proposing and studying the

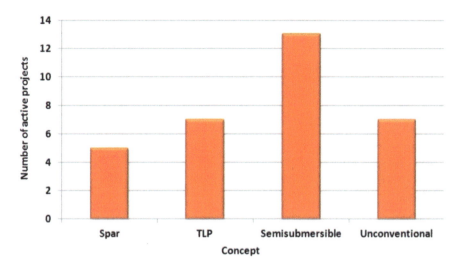

Fig. 4.2 Floating offshore wind projects sorted based on the support-structure concept. *TLP* tension-leg platform

Table 4.1 Floating offshore wind turbines (examples of MW turbines)

Name	Country	Installed capacity (MW)	Turbine model	Structure
WindFloat	Portugal	2	Vestas V80-2.0 MW	Semisubmersible (3 columns)
Fukushima	Japan	2	Subaru 80/2.0	Semisubmersible (4 columns)
Hywind	Norway	2.3	Siemens SWT-2.3-82 VS	Spar

Fig. 4.3 Hywind, spar-type wind turbine. (Courtesy of Vines 2009; this file is licensed under the Creative Commons Attribution 3.0 Unported license)

Table 4.2 Hywind project characteristics. (statoil.com 2012; Lore 2011; nexans.com; Technip 2009)

Status	Commissioned in 2009
Location	Karmøy Island, Norway (installed in the North Sea)
Distance from shore	10 km
Testing period	Originally for 2 years, but it is extended
Installer turbine/spar/cable	Aker solutions/Technip/Nexans
Turbine size	2.3 MW
Turbine weight	138 t
Turbine height	65 m
Rotor diameter	82.4 m
Draft hull	100 m
Displacement	5300 m^3
Diameter at water line	6 m
Diam. submerged body	8.3 m
Water depths	200 m
Mooring	Three sets of lines

feasibility of floating wind turbines and their behavior under wave and wind actions had been made by scientists around the world before Hywind installation. Hywind is a spar-type floating wind turbine using a spar platform as the base. The platform is moored to the seabed using catenary mooring. Three mooring sets have been used for station keeping. The turbine tower is located at the top of the spar. In principle, spar is a circular cylinder, a buoy, which is ballasted using water and rock.

Spar concept has been widely used in offshore oil/gas technology. It is a proven concept working well. Technip has good experience in design and construction of spar platforms. Technip was the main contractor for the structural parts of Hywind. Technip designed, constructed, and delivered the spar. Turbine manufacturer was Siemens wind power. Siemens SWT-2.3-82 VS turbine with rated power of 2.3 MW was bought and mounted over the spar. Turbine characteristics are listed in Table 4.4. The known wind and marine technologies were combined in a new setting for floating wind turbines. This was a good attempt to open up the possibility for capturing wind energy in deep water environments. Statoil, a leading offshore oil/gas operator, used its valuable experience to develop Hywind as the first full scale offshore floating wind turbine. Hywind demo project aimed to discover the challenges for this new technology to enhance the development of knowledge needed.

The floating part consists of a steel cylinder filled with a ballast of water and rocks. Draft of spar is around 100 m, and the spar is moored using spread catenary mooring. The intention of the demo project is to find the behavior of the spar-type wind turbine under the action of wave and wind in real life. Hence, survivability and structural strength are monitored. Power generation and its deficit due to the action of waves and platform motions are the other aspects of this project. The next generation of Hywind may have improvements based on the lessons gained. The goal

Table 4.3 Annual power production of Hywind. (The-Hywind-O&M-Team 2012)

Year	Annual production (GWh)	Capacity factor
2012	7.5	0.37
2011	10.1	0.50
2010	7.4	0.36

is to commercialize the concept by reducing costs, so that floating wind power can compete in the energy market (statoil.com 2009). The Hywind concept is designed for deep water; Japan, the USA, and the UK can be nominated regions, which have deep water areas (Next step can be a wind park of 3–5 turbines). Hywind Scotland pilot park project consists of 30 MW with five turbines each of which will be up to 5–6 MW in size. The project is planned to be completed by 2016–2017 (xodus-group.com 2013).

Statoil invested around NOK 400 million in the construction and further development of the pilot project and in research and development related to the wind turbine concept. After more than 2 years of operation, the concept had been verified, and its performance was beyond expectations. With few operational challenges, excellent production output, and well-functioning technical systems, the Hywind concept could revolutionize the future of offshore wind (statoil.com 2009).

Annual estimated production of Hywind is 7.9 GWh per year which corresponds to capacity factor of 39; $CF = \dfrac{7.9\ GWh}{2.3\ MW \times 365 \times 24\ h} = 0.39$. The Hywind production in the past years is listed in Table 4.3.

Operating voltage level is 22 kV; Nexans 24 kV exporting cable with the length of 13.6 km is used and power frequency at onshore is 50 Hz (nexans.com). Turbine characteristics are listed in Table 4.4. The tower is mounted at the top of the spar platform by bolting at the flange. The spar has a deep draft and it is stabilized by ballasting. The mooring lines are attached to fairleads below mean water surface and keep the structure. The mooring lines are catenary and spread around the spar. The other end of the mooring lines is connected to seabed. This setting allows slowly varying motions such as slowly varying surge and sway responses. More discussion regarding spar-type wind turbines aspects are given later in this chapter.

4.4   WindFloat Project

WindFloat is a semisubmersible type floating offshore wind turbine. The name of the project is Demowfloat, which is supported by FP7 of the European Commission and gathers 12 entities of five different countries. The goal is to test and monitor the performance of WindFloat. The WindFloat concept is designed for deep water more than 40–50 m for harnessing offshore wind power. The prototype project is located 6 km offshore Portugal, at a depth of about 42 m (demowfloat.eu).

Table 4.4 Hywind turbine characteristics (siemens.com 2009b; Lorc 2011a; siemens.com 2009)

Turbine model	Siemens SWT-2.3-82 VS
Rated power	2.3 MW
Wind class	IEC IA
Cut-in wind speed	4 m/s
Rated wind speed	13.5 m/s
Cut-out wind speed	25 m/s
Rotor type	3-bladed, horizontal axis
Rotor position	Upwind
Rotor diameter	82.4 m
Rotor area	5333 m^2
Rotor speed (minimum)	6 rpm
Rotor speed (rated)	18 rpm
Rotor weight (incl. hub)	55 t
Hub height (above MSL)	65 m
Blade tip speed (rated)	77.66 m/s
Blade length	40 m
Blade root chord	3.1 m
Blade tip chord	0.8 m
Power regulation	Pitch regulated with variable speed (hydraulic)
Drivetrain type	High speed
Main bearing	Spherical roller bearing
Gearbox ratio	0.010989
Gearbox stages	3 planetary stages, 1 helical stage
Gearbox lubrication	Splash/forced lubrication
Generator type	Asynchronous
Generator rated power	2300 kW
Power frequency	50 Hz
Turbine voltage level	690 V
Transformer voltage level	22 kV
Nacelle weight	83 t

The turbine is commissioned and is already in operation since December 2011. WindFloat is a semisubmersible type wind turbine with three main columns. The platform consists of three columns that provide buoyancy to support the turbine. The columns are separated from each other to provide stability by increasing the metacentric height. The columns are connected by braces to each other, and the turbine is located above one of the columns. Some characteristics of the installed turbine are mentioned below:

The aim of the Demowfloat project is to study a semisubmersible floating wind turbine called WindFloat, see Fig. 4.4. The structural integrity and the performance

Fig. 4.4 WindFloat, a semisubmersible type floating offshore wind turbine foundation operating at the rated capacity (2 MW) approximately 5 km offshore of Agucadoura, Portugal. (Source: commons.wikimedia.org (Untrakdrover 2012); this file is licensed under the Creative Commons Attribution 3.0 Unported license)

of the system are monitored. The operationality, maintainability, reliability, platform accessibility, feasible grid integration on a modular basis, and other aspects are considered to assess the cost of produced energy from this floating offshore wind turbine. This is the start of a wind farm based on semisubmersible type support structure (demowfloat.eu). The wind turbine is the commercial Vestas model V80-2.0 MW, which has already been tested for onshore and offshore environments. The structure is anchored to the seabed by four mooring lines made of conventional components (Table 4.5).

Table 4.5 Characteristics of WindFloat. (Vidigal 2012)

Displacement	2750 t
Ballast	1200 t
Draft	13.7 m
Freeboard	9.5 m
Tower	54 m
Nacelle height above MWS	63.5 m
Distance between columns	38 m
Column diameter	8 m
Rated power	2 MW
Rotor diameter	80 m

WindFloat is originated in a company called PrinciplePower (principlepower-inc.com). This concept is suitable for intermediate (>40 m) and deep-water depth offshore wind energy market. The term "deep" in offshore wind refers to water depth more than 100 m. However, traditionally, in offshore oil/gas business "deep" refers to much more depth, e.g., 300 m.

In WindFloat, damping plates at the bottom of the columns are set to damp motions, especially in heave motion. However, the capability and performance of such design should be investigated, thoroughly, keeping in mind the added mass effects of the damping plates. Heavy lifting is avoided in WindFloat as it is commissioned before transportation. This is the advantage of semisubmersible wind turbine and they could be wet towed to the site.

WindFloat has a closed-loop hull trim system that mitigates the effect of mean wind-induced thrust forces. Hence, it has an active ballasting together with static ballasting. This system helps optimizing energy production considering the changes in wind velocity and direction (principlepowerinc.com). However, the performance and reliability of such controlled system should be checked as well.

Four mooring lines are used, two are attached to the column which is supporting the turbine, and the other two are attached to the other column. WindFloat employs conventional mooring components such as chain and polyester lines to minimize the cost and complexity. Through the use of pre-laid drag-embedded anchors, site preparation is minimized (principlepowerinc.com).

WindFloat project in Portugal has three phases, 2, 27, and 150 MW. The first phase is accomplished, and the turbine is operating (EDP 2012). The second phase is a pre-commercial phase consisting of 3–5 turbines (Maciel 2010). The third commercial stage consists of approximately 30 wind turbines with the rotor diameter of 120–150 m and total height of 160–175 m (4C-Offshore, 2013).

4.5   Fukushima Project

After nuclear disaster in Japan due to Tsunami in 2011, it was decided to shutdown the nuclear power plants. The government supported a consortium called Fukushima offshore wind to develop offshore wind power based on floating support structures, see Table 4.6.

Fukushima offshore wind consortium, see Table 4.7, which consists of Marubeni Corporation (project integrator), the University of Tokyo (technical advisor), Mitsubishi Corporation, Mitsubishi Heavy Industries, Japan Marine United Corporation, Mitsui Engineering & Shipbuilding, Nippon Steel & Sumitomo Metal Corporation Ltd., Hitachi Ltd., Furukawa Electric Co. Ltd., Shimizu Corporation and Mizuho information & Research, is proceeding with Fukushima floating offshore wind farm demonstration project (Fukushima FORWARD) funded by the Ministry of Economy, Trade and Industry since 2012 (marubeni.com 2013).

Fukushima project has two stages (fukushima-forward.jp, Fukushima FORWARD) in which three floating wind turbines (two semisubmersibles and one spar)

Table 4.6 Fukushima FORWARD project phases. (marubeni.com 2013)

Fukushima FORWARD	Facility name	Scale	Form	Floating form
Phase I (2011–2013)	Floating Substation "Fukushima Kizuna"	25 MVA	Substation (with 66 kV high-voltage undersea cable)	Advanced spar
	Floating Wind Turbine "Fukushima Mirai"	2 MW	Downwind type	4 column Semisubmersible
Phase II (2014–2015)	Large Floating Wind Turbine "Fukushima Shimpuu"	7 MW	Oil pressure drive type (Hydraulic turbine)	3 column Semisubmersible
	Large Floating Wind Turbine	7 MW	Oil pressure drive type (Hydraulic turbine)	Advanced spar

Table 4.7 The role of each member in the Fukushima FORWARD consortium (fukushima-forward.jp, Fukushima FORWARD)

FORWARD member	Main role
Marubeni Corporation	Feasibility study, approval and licensing, O & M, collaboration with fishery industry
The University of Tokyo	Metocean measurement and prediction technology, marine navigation safety, public relation
Mitsubishi Corporation	Coordination for grid integration, environmental impact assessment
Mitsubishi Heavy industries Ltd.	V-shape semisubmersible (7 MW)
Japan Marine United Corporation	Advanced spar-type floating substation
Mitsui Engineering & Shipbuilding Co., Ltd.	Compact semisubmersible (2 MW)
Nippon Steel & Sumitomo Metal	Advanced steel material
Hitachi Ltd.	Floating substation
Furukawa Electric Co., Ltd.	Large capacity undersea cable
Shimizu Corporation	Pre-survey of ocean area, construction technology
Mizuho Information & Research institute, Inc	Documentation, committee operation

and one spar-type floating power substation will be installed off the coast of Fukushima. The first phase of the project consists of one 2 MW semisubmersible-type floating wind turbine, the world's first 25 MVA spar-type floating substation, and 66 kV undersea cables.

In November 2013, it was announced that the first phase of the project has been successfully accomplished. Installation of 2 MW downwind-type semisubmersible wind turbine, 25 MVA spar-type power substation, 66 kV extra-high voltage undersea cables, and the dynamic cable was successfully completed (marubeni.

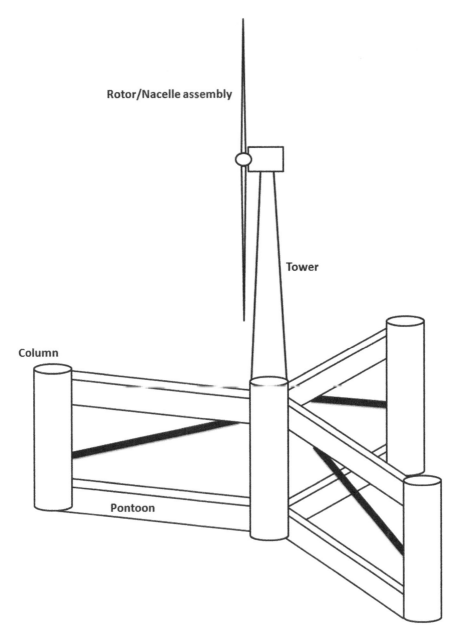

Fig. 4.5 Schematic layout of 2 MW Fukushima semisubmersible wind turbine (fukushima-forward.jp, Fukushima FORWARD)

com, 2013). The installed 2 MW turbine is a 4-column semisubmersible in which the central column carries the turbine, see Fig. 4.5. It is so-called "compact" semisubmersible in which braces are increasing the structural integrity. The substation is based on an "advanced" spar floater, see Fig. 4.6.

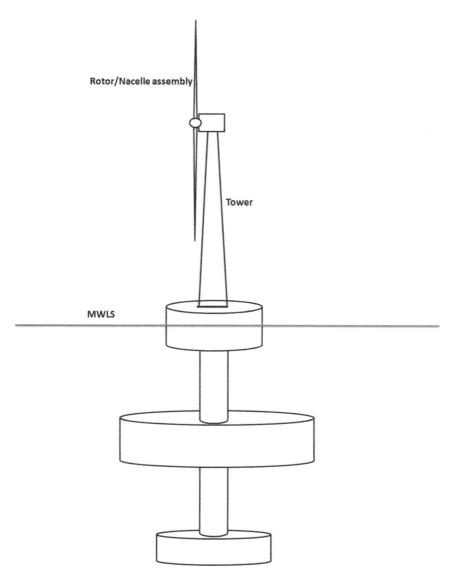

Fig. 4.6 Schematic layout of Fukushima advanced spar-type floating wind turbine. Mean Water Level Surface (*MWLS*) (fukushima-forward.jp, Fukushima FORWARD)

One aspect of having the floating substation is to collect the meteorological and hydrographic data. The metocean measurement system considers the floater motion compensation. The performance of the semisubmersible turbine is studied to evaluate safety, reliability, and the economic potential of the offshore floating wind farm. In addition, the project aims at establishing the method of operation and maintenance of the offshore floating wind farm at the same time.

In the second phase, the two units of 7 MW oil pressure drive-type floating wind turbines will be installed within the fiscal year 2014. One of the turbines is supported by the spar platform, and the other one is based on semisubmersible. An advanced spar and a braceless semisubmersible (V-shaped with three columns) will be applied. Hence, the total capacity of the offshore floating wind farm at Fukushima will be 16 MW all together (two 7 MW and one 2 MW), making it the world's biggest floating offshore wind farm (fukushima-forward.jp, Fukushima FORWARD)(fukushima-forward.jp, Fukushima FORWARD). In compact semisubmersible, the mooring system has six pieces catenary mooring lines, and in V-shaped semisubmersible, the mooring system has eight pieces catenary mooring lines.

4.6 Floating Wind Turbine Concepts

Floating wind turbines are supported by floating structures and, hence, have 6 degrees of freedom, which can be excited by wave, wind, and ocean current loads. The entire system should be moored and stabilized using mooring lines, ballasting, etc. They are relatively huge structures varying 5000–10,000 t for a 2–5 MW unit.

The base cases are spar, tension-leg platform (TLP), ship shaped (e.g., barge) and semisubmersible. However, hybrid concepts and modified concepts can be considered as well. As mentioned before, only a few floating wind turbines are installed. One is the Hywind in Norway, fitted with a turbine from Siemens. Another is the Windfloat, installed off the coast of Portugal, with a Vestas turbine. The most recent is the Fukushima semisubmersible wind turbine. Few scale models such as the Blue H in Italy and Sway in Norway are constructed as well.

Any kind of stabilized and moored floating body can be considered as a base structure for a wind turbine. Brainstorming in several joint projects had been carried out, and different concepts and designs for floating wind turbines were introduced. The idea is to have cost-effective solutions capable of competing with bottom-fixed offshore wind turbines in the short term, and decreasing the need of subsidies for offshore wind business in a longer-time scale and compete with hydrocarbon energies, such as oil and gas.

Offshore wind is a good resource providing stronger and steadier winds, hence, increasing the annual production. Due to technical challenges and cost issues, special concepts are feasible in such water depths. Different proven support structures from the offshore oil/gas industry can be the starting point for feasible floating offshore wind designs. They need to be modified and tailor-made to suit the requirements needed in the wind industry. The unnecessary issues, e.g. high reliability factor (low risk) applied in oil business, should be customized as well. In the following, some of the basic concepts are discussed.

4.7 Semisubmersible Offshore Wind Turbine

Offshore wind projects that applied semisubmersible as the base floater are listed in Table 4.8. Semisubmersible type floating wind turbines can be installed and commissioned near shore and transported afloat to offshore site. This is one of the key

Table 4.8 Semisubmersible floating offshore wind projects

Company, country	Offshore site	Water depth	Status	Future development	Website/references
WindFloat Principle power (USA)	Portugal	50 m	2 MW operating since 2011	More units may appear	http://www.principlepowerinc.com (demowfloat.eu) (principlepowerinc.com)
Compact semisubmersible Mitsui Fukushima FORWARD (Japan)	Japan	–	2 MW wind turbine commissioned in 2013	–	https://www.mes.co.jp/english/ (fukushima-forward.jp, Fukushima FORWARD)
DeepCWind(USA)	USA	–	Scaled prototype (20 kW) launched in 2013	6 MW appears in 2016	http://www.deepcwind.org/
HiPR Wind (EU)	Spain	80 m	Design phase	1.5 MW is planned	http://www.hiprwind.eu/
Windflo Nass et wind (France)	France	50 m	Design phase	MW scale	http://www.nass-et-wind.com/ (4coffshore.com 2013)
Vertiwind Technip (France)	France	50 m	Design phase	2 MW planned	http://www.technip.com/en
Mitsubishi Fukushima FORWARD (Japan)	Japan	–	–	7 MW planned	http://www.mhi.co.jp/ (fukushima-forward.jp, Fukushima FORWARD)
INFLOW Technip (France)	France	–	–	2 MW	http://www.technip.com/en

Table 4.8 (continued)

Company, country	Offshore site	Water depth	Status	Future development	Website/references
GustoMSC Trifloater (Netherlands)	–	50 m	Tank test	–	http://www.gustomsc.com/
Hitachi (Japan)	Japan	–	Cooperation with Statoil	–	http://www.hitachizosen.co.jp/
Shimizu (Japan)	Japan	25 m	Design phase	–	http://www.shimz.co.jp/
WEMU (EU)	Russia	5 m	Scaled prototype testing	–	http://www.dvfu.ru/
WindSea (EU)	Norway	25 m	Tank tests completed	–	http://www.windsea.no/about-windsea/

advantages of this concept. Semisubmersible projects with and without braces are introduced in the market. Deployment of braces limits the fatigue life and affects the design. Fukushima phase I applied semisubmersible with braces. However, in phase II of Fukushima project, braceless semisubmersible is planned to be used as a base structure for a 7 MW turbine. Figure 4.7 illustrates the schematic layout of a semisubmersible wind turbine.

Semisubmersible floaters are gaining stability by spreading the water surface area. They usually consist of 3–4 slender columns that are connected to each other by pontoons and braces (in offshore oil/gas, they can have more columns depending on the design). The restoring moments depend on the surface area of each column and the distance between them $(\propto AL^2)$. The distance between the center of gravity (COG) and the center of buoyancy (COB) can increase the restoring moments as well (if the center of mass is below). However, for semisubmersibles, the main contributor in stability is the arrangement of the columns and the surface area of them. Increasing the surface area means more hydrodynamic forces and consequently more structural stiffness needed to cope with the loads. The increase of the distance between columns requires more stiffening of braces and pontoons. In Fig. 4.8, an example of braceless concept, V-shaped semisubmersible, is shown. The pontoons are directly connected to columns.

A semisubmersible offsore wind is stable in heave due to Archimedes law. The total weight of the structure is in balance with the buoyancy force. If the platform moves downward, the added volume of submerged part of columns applies a buoyancy force upward and tries to return the structure back to its initial condition. Pitch and roll motions are stabilized by the action of restoring moments. The restoring moments are discussed above. The yaw, surge, and sway need mooring line actions to be stabilized. The mooring keeps the system stable while allowing some

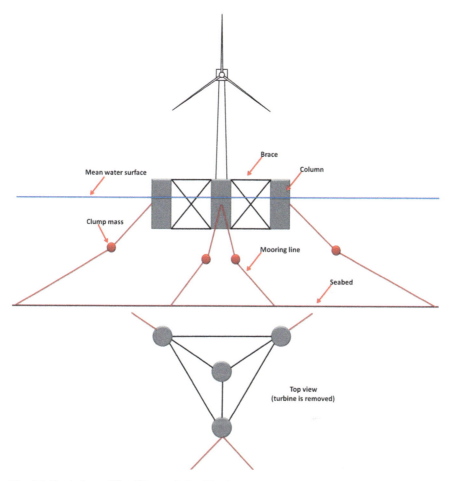

Fig. 4.7 Semisubmersible offshore wind turbine layout

freedom for slowly varying motions. Surge and sway have natural frequencies (e.g., 0.05 rad/sec) much lower than the wave frequency. Yaw motion can have larger Eigen-mode frequency, still below the wave frequency region.

4.8 Tension-Leg Platform (TLP) Offshore Wind Turbine

Tension-leg platform (TLP) offshore wind projects are listed in Table 4.9. Schematic layout of a TLP wind turbine is shown in Fig. 4.9. A TLP is stabilized using the tension forces in the tendons. The tension legs compensate the force difference between buoyancy and total weight. The ratio between the total tension in legs and

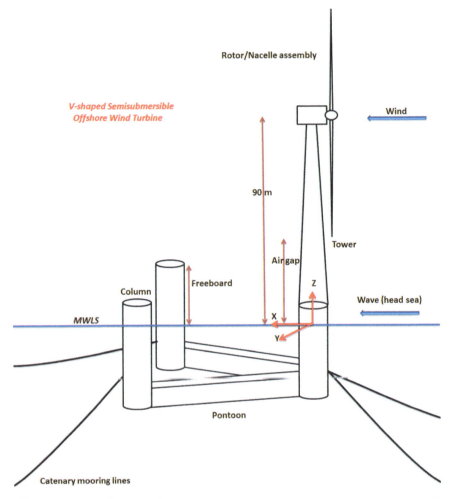

Fig. 4.8 Example of the braceless concept, V-shaped semisubmersible wind turbine. *MWLS* Mean Water Level Surface

the total weight is in the order of 25 %. Installation of such a system can be a challenge. One way is to ballast the system prior to transport/installation and de-ballast it again prior to installation of tension legs. However, deep study of the stability during all these phases is necessary. It is likely that the metacentric height of such a system is negative. This means if the tendons are removed, the structure is not stable anymore. TLPs have small motions resembling as they are fixed. This can be an advantage to gain more smooth electric power. However, to keep a structure so-stiff means resisting against the hydro-aero-dynamic loads. Also, the appearance of

Table 4.9 Tension-leg platform wind projects

Company, country	Offshore site	Water depth	Status	Future development	Website/ references
GICON SOF (Germany)	Germany	20 m	Design phase	–	http://www.gicon.de/
BlueH (Netherlands)	Italy	50 m	80 kW installed in 2008	MW scale planned	http://www.bluehgroup.com/
[a]Mitsui (Japan)	Japan	60 m	–	–	https://www.mes.co.jp/
Ocean breeze (EU)	UK	60 m	–	–	http://www.xanthusenergy.com/
[a]Nautica (USA)	USA	–	–	–	http://www.nauticawindpower.com/
Glosten Pelastar (USA)	UK	65 m	Design phase	MW scale planned	http://www.pelastarwind.com/
Iberdrola (Spain)	UK	–	Model testing in 2012	–	www.iberdrola.es

[a] Hybrid TLP/semisubmersible

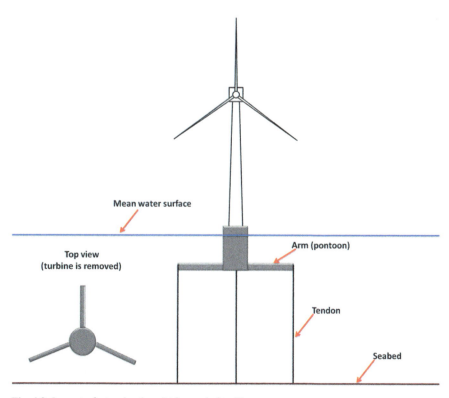

Fig. 4.9 Layout of a tension-leg platform wind turbine

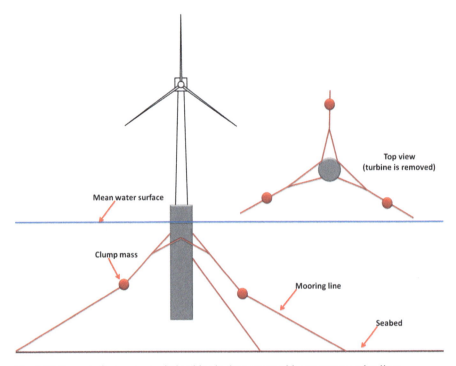

Fig. 4.10 Layout of a spar-type wind turbine in deep water with catenary mooring lines

higher Eigen-modes in TLP concepts needs care for being out of excitation by rotor dynamics and tower elastic modes.

A TLP wind turbine consists of a central column that is usually slender to reduce hydrodynamic loads. The tower and rotor/nacelle assembly is mounted at the top of this column. Tendons need to be attached with a distance to the central column. This provides restoring moments. Stiff arm like a bar is an option otherwise pontoons can be applied depending on the design. Use of pontoon helps to reduce the size of the central column and consequently to reduce the hydro loads. Meanwhile, the ballasting can be applied at lower compartments to increase the stability. Suction or gravity anchoring may be applied to fix the tendons at the seabed.

4.9   Spar Offshore Wind Turbine

Figure 4.10 shows a catenary moored spar-type wind turbine in deep water. The offshore wind projects using spar floater are listed in Table 4.10. In general, spar platform is a circular cylinder which is ballasted using water/metal/concrete at lower

Table 4.10 Spar offshore wind projects

Name company	Offshore site	Water depth	Status	Future development	Website/references
Hywind Statoil (Norway)	Norway	200 m	2.3 MW operating since 2009	Farms based on 5 MW turbines	http://www.statoil.com/ (xodusgroup.com 2013) (statoil.com 2009)
Sea Twirl (Sweden)	Sweden	–	Scale prototype tested	–	http://seatwirl.com/
Nagasaki/ Goto-hybrid-spar Toda Cooperation (Japan)	Japan	100 m	100 kW launched in 2012	2 MW is commissioned 2013 by replacing the 100 kW, Larger turbine appears 2016	http://www.kyoto-u.ac.jp/en (4coffshore.co, 2013) (japanfs.org, 2009) (offshorewind.biz 2013)
Fukushima FORWARD (Japan)	Japan	50 m	Full scale for substation (not a wind turbine)	7 MW wind turbine appears	http://www.jmuc.co.jp/en/ (fukushima-forward.jp, Fukushima FORWARD)
Sway (Norway)[a]	Norway	55 m	Scaled prototype testing	–	http://www.sway.no/

[a] Sway: a hybrid Spar/TLP concept

compartments. This lowers the COG and increases the distance between the COB and COG. High metacentric height (GM) helps to increase the stability of the structure. Tower and rotor/nacelle assembly are put at the top of the spar.

The restoring moments in pitch and roll motions are directly linked to GM. The heave motion restoring comes from the surface area. The surge/sway and yaw need stiffness of mooring lines. There is no hydrodynamic excitation for yaw motion; hence, for regular oil/gas platform-spread mooring lines around the platform attached to fairleads are adequate. However, for a floating wind turbine, the wind loads introduce yaw moment, which should be taken by mooring lines. Due to the slender shape of spar platforms, if the mooring lines are directly attached to fairlead, a small restoring moment in yaw is resulted. Hence, the mooring lines should be attached with an arm, e.g., attached to a horizontal bar. Another option is having delta lines. Delta configuration results in proper yaw-restoring-moments if clump mass is used to increase the tension.

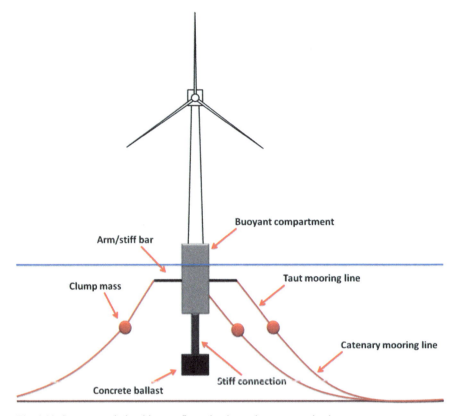

Fig. 4.11 Spar-type wind turbine configuration in moderate water depth

Spar can be implemented in moderate water depth, e.g., 100–150 m if proper considerations are made in design regarding the mooring lines and mass/buoyancy configuration. The limitation of using spar in shallower water is highly dependent on the size of the turbine. Figure 4.11 shows an example for a spar-type wind turbine configuration in moderate water depth.

4.10 Unconventional Floating Wind Turbine Concepts

There are some offshore wind projects in which alternative floaters and arrangements compared to traditional concepts have been used, see Table 4.11. Different concepts based on combining the regular concepts have been introduced. These

Table 4.11 Alternative offshore wind projects

Name company	Offshore site	Type/platform	Water depth	Status	Website references
DIWET (France)	–	Floater	–	–	(4coffshore.com 2013)
Hexicon (EU)	–	Floating wind turbines array	26 m	Design phase	http://www.hexicon.eu/
SKWID MODEC (Japan)	Japan	Floating wind and current hybrid power generation Supported by moored buoy	–	Onshore test for wind part in 2013	http://www.modec.com/fps/skwid/
IDEOL (France)	France	Ring-shape surface floater	35 m	–	http://www.ideol-offshore.com/en
Poseidon (Denmark)	Denmark	Combined wave and wind, ship-shaped semisubmersible	40 m	140 kW from waves and 33 kW from wind	http://www.floatingpower-plant.com/
Pelagic (Norway)	Norway	Combined wave and wind, semisubmersible	–	–	http://www.pelagicpower.no/
Wind Lens Kyushu University (Japan)	Japan	Combined wind and solar, semisubmersible	–	6 kW commissioned offshore	http://www.riam.kyushu-u.ac.jp/

hybrid concepts try to gather the advantages of the basic concepts to cope with the challenges and requirements of offshore wind technology. Hybrid marine platforms combining wave energy converters, wind turbines, and ocean current turbines are rapidly appearing. Synergy effects, which increase the power and lower the cost, are the main points of such ideas. In Fig. 4.12, an example of combining concepts is shown. In this proposed concept, a semisubmersible offshore wind turbine applies tension legs. Hybrid marine platforms combining the wave and wind power devices are discussed later in the current book.

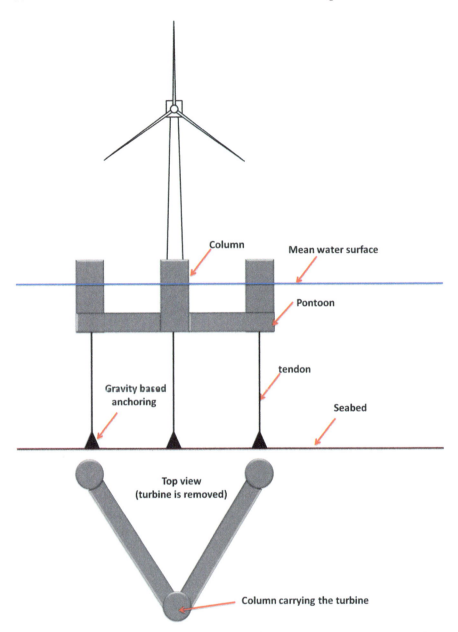

Fig. 4.12 An innovative example of floating wind turbine based on a combination of semisubmersible and tension-leg platform

Bibliography

4C-Offshore. (2013). WindFloat. http://www.4coffshore.com/windfarms/windfloat---phase-1-portugal-pt01.html. Accessed March 2014.

4coffshore.co. (2013). GOTO FOWT floating offshore wind turbine—100 kW. http://www.4coffshore.com/windfarms/goto-fowt-floating-offshore-wind-turbine----100kw-japan-jp15.html. Accessed March 2014.

4coffshore.com. (2013). Anholt. http://www.4coffshore.com/windfarms/anholt-denmark-dk13.html. Accessed March 2014.

4coffshore.com. (2013). DIWET Deepwater Innovative Wind Energy Technology. http://www.4coffshore.com/windfarms/diwet-deepwater-innovative-wind-energy-technology-france-fr06.html. Accessed March 2014.

4coffshore.com. (2013). Robin Rigg. http://www.4coffshore.com/windfarms/robin-rigg-united-kingdom-uk20.html. Accessed March 2014.

4Coffshore.com. (2013). Tripile support structures. http://www.4coffshore.com/windfarms/tripile-support-structures-aid272.html. Accessed March 2014.

4coffshore.com. (2013). West of Duddon Sands. http://www.4coffshore.com/windfarms/west-of-duddon-sands-united-kingdom-uk33.html. Accessed March 2014.

4coffshore.com. (2013). WINFLO Wind turbine with INnovative design for Floating Lightweight Offshore. http://www.4coffshore.com/windfarms/winflo-wind-turbine-with-innovative-design-for-floating-lightweight-offshore-france-fr07.html. Accessed March 2014.

EDP. (2012). The WindFloat Project. http://ec.europa.eu/maritimeaffairs/policy/sea_basins/atlantic_ocean/atlanticforum/events/brest/presentations/forum_brest_maciel_en.pdf.

japanfs.org. (2009). World's first hybrid spar-type platform for floating wind turbine succeeds in demonstration test. http://www.japanfs.org/en/news/archives/news_id029511.html. Accessed March 2014.

Lorc. (2011a). Hywind demonstration offshore wind farm. http://www.lorc.dk/offshore-wind-farms-map/hywind-demonstration. Accessed March 2014.

LORC. (2011). Tripile—Three monopiles in one. http://www.lorc.dk/offshore-wind/foundations/tripiles. Accessed March 2014.

marubeni.com. (2013, November 11). Fukushima recovery, experimental offshore floating wind farm project. http://www.marubeni.com/news/2013/release/fukushima_e.pdf. Accessed March 2014.

siemens.com. (2009b). Siemens Wind Turbine SWT-2.3-82 VS. http://www.energy.siemens.com/hq/pool/hq/power-generation/wind-power/E50001-W310-A123-X-4A00_WS_SWT-2.3-82%20VS_US.pdf. Accessed March 2014.

statoil.com. (2009). Hywind—The world's first full-scale floating wind turbine. http://www.statoil.com/en/technologyinnovation/newenergy/renewablepowerproduction/offshore/hywind/pages/hywindputtingwindpowertothetest.aspx. Accessed March 2014.

statoil.com. (2012). Hywind. http://www.statoil.com/no/technologyinnovation/newenergy/renewablepowerproduction/offshore/hywind/pages/hywindputtingwindpowertothetest.aspx. Accessed March 2014.

Technip. (2009). Technip joins StatoilHydro in announcing the inauguration of the world's first full-scale floating wind turbine. http://www.technip.com/en/press/technip-joins-statoilhydro-announcing-inauguration-worlds-first-full-scale-floating-wind-turbi. Accessed March 2014.

The-Hywind-O & M-Team. (2012). Hywind: Two years in operation. http://www.sintef.no/project/Deepwind%202012/Deepwind%20presentations%202012/D/Trollnes_S.pdf. Accessed March 2014.

Untrakdrover. (2012). Agucadoura WindFloat Prototype. http://commons.wikimedia.org/wiki/File:Agucadoura_WindFloat_Prototype.jpg. Accessed March 2014.

Vidigal, A. (2012). WindFloat: Quase há um ano no Mar. http://www.antoniovidigal.com/drupal/cd/WindFloat-Quase-h%C3%A1-um-ano-no-Mar. Accessed March 2014.

Vines, J. (2009). Hywind havvindmølle. http://commons.wikimedia.org/wiki/File:Hywind_havvindm%C3%B8lle.JPG. Accessed March 2014.

xodusgroup.com. (2013). Hywind Scotland Pilot Park Project. Edinurgh: A-100142-S00-REPT-001.

Chapter 5
Wave Energy Converters

5.1 Introduction

Wave power presents as the movements of water particles close to the ocean surface. The energy intensity depends on the height and frequency of waves. A large amount of wave power in random sea motivates us to think of using ocean wave energy for generating electricity. We use wave energy converters (WECs) to change potential kinematical energy of sea waves to electrical energy. Figure 5.1 shows an example of WECs. Waves in the ocean have a vast amount of renewable energy. This makes the ocean a renewable source of power in the order of terawatts (TW). The global power resource represented by waves that hits all coasts worldwide, has been estimated to be in the order of 1 TW (Falnes 2002).

Sun creates an uneven distribution of pressure regions in the earth's atmosphere. The air starts to move between pressure fields and, hence, the wind blows over seas and oceans. Boundary layers are created over the water surface, and water particles start to move; thus, waves are formed. Wind-generated waves and traveling waves with long periods (swell) have a high density of energy compared to wind energy. Wave energy provides 15–20 times more available energy per square meter than either wind or solar (Muetze and Vining 2006). The reason is that the waves are created by progressive transfer of energy from wind and the water carries that energy (note: the water density is roughly 800 times higher than air). Scientists have studied this dense energy for decades and several concepts have been introduced in the wave energy field. However, a limited number of them could survive for further developments and maturity.

Wind energy had a good growth over the past years. In the same way, onshore wind technology has developed, and offshore wind technology based on fixed structures is rapidly developing. However, the wave energy did not have such growth. In 2002, 340 patents were existed regarding wave energy conversion, with the first known patent dating back to 1799. Several research programs were started mainly by the European countries, such as the UK, Portugal, Sweden, Denmark, Ireland, and Norway, 40 years ago when the oil crisis started (Engstrom 2011).

© Springer International Publishing Switzerland 2014
M. Karimirad, *Offshore Energy Structures*, DOI 10.1007/978-3-319-12175-8_5

Fig. 5.1 Example of wave energy converters (WECs). (Courtesy of Ingvald Straume (Straume 2014). This file is made available under the Creative Commons CC0 1.0 Universal Public Domain Dedication)

Following the oil crisis in 1973, many researchers at universities and other institutions took up the subject of wave energy (Falnes 2002). The topic of wave energy conversion came to public attention in a paper by Steven Salter published in Nature in 1974 (McCormick 2007). The UK, Japan, Norway, Sweden, France, and the USA were the pioneers. For more information, refer to Ocean Wave Energy Conversion by McCormick (McCormick 2007).

During the early 1980s, when the petroleum price declined, wave-energy funding drastically reduced (Falnes 2007). However, the wave energy activities have been strongly continued. As an example (McCormick 2007), Clive O. J. Grove-Palmer, the program manager of the UK Wave Energy Technology Support Unit said: "The wave energy community does not give up easily and the prize will not be easily won, however."

Commercially, the economics of wave energy is becoming more attractive. Grater electricity costs, possible carbon taxes (as high as US$ 50/t), and increasing subsides (up to € 0.24/kWh, International Energy Agency, IEA) will make even the early generation of wave devices more economic (Tedd 2007).

However, there is no mature commercial WEC at the moment. Hence, more research and studies considering numerical, experimental, and demo projects are needed to fill the technical gaps and introduce mature and cost-optimized concepts.

Day to day, the demand of energy is increasing and environmental issues such as pollution and global warming urge scientists to act. In the past decade, activities have been increased and several wave power concepts have had offshore testing. Many concepts are also advancing into the first commercial plants and demonstration projects such as Pelamis (pelamiswave.com 2013a), Seabased (seabased.com 2013), and Wavegen (Seed 2012).

There are some challenges for wave power that are coming from the nature of ocean waves and have decelerated the enhancement of this technology. First of all, the ocean waves are stochastic with irregularity in amplitude with respect to time. The water surface elevation and frequency are continuously changing. Hence, the energy magnitude (which is related to wave amplitude and frequency) varies rapidly, e.g., from 0 to 200 kW/m. The average wave power is considered a good first indicator of how much energy is available at a particular site. This is the available energy and just a portion of it can be converted into electrical energy. An array of wave energy convertor is needed to compensate the high fluctuation in the output power. Otherwise, special electrical system is required. When an array of convertors are considered, it is necessary to focus more on the possible conflicts with fisheries, shipping, environmental, and cultural issues.

Most of the available metocean data are related to offshore oil/gas technology. The energy potential of a site is related to energy transportation by wave which is connected to sea states. Some part of the available energy can be converted into electricity. It is highly important to have a correct scatter diagram and metocean data in order to design the structure and optimize the WEC with respect to the power take off (PTO) system. More refined metocean data are needed to help maturing and advancing this technology.

The next challenge is the high ratio between maximum available power and the average value. The ocean wave power is highly skewed. The design of electrical components, structural parts, and mooring lines that can handle the maximum loads results in high cost. The next problem is survival conditions and how the system should be parked (shutdown) in a way to minimize the loads. As it is discussed earlier, ocean waves have high energy density; hence, to decrease the cost of electricity produced from WEC, a high-power capture ratio is required.

Similar to other floating structures, the WEC is a dynamical system which its responses to ocean stochastic loads should be considered. Motions, deflections, stresses, and strains of the structure have a great role in design and developments of the WEC. A WEC is designed to take-off power from the sea. This role cannot be correctly done unless the structure survives in all sea states. For making more cost efficient power, sufficient knowledge about motions in real sea is necessary.

5.2   Wave Energy Resources

As it is mentioned above, the waves are generated due to transfer of wind energy to water on ocean surfaces. So, great waves are generated when strong winds blow over long distances. A long fetch is an essential factor for a good offshore wave resource. The power or energy flux (P) transmitted by a regular wave per unit width can be written as follows (Cornett 2008):

$$P = \frac{1}{8} \rho_{water} g C_g H^2, \tag{5.1}$$

where ρ_{water} is the water density, H is the wave height and C_g is the group velocity defined by:

$$C_g = \frac{1}{2} \left[1 + \frac{2kh}{\sin h(2kh)} \right] \frac{\lambda}{T}, \tag{5.2}$$

where h is the mean water depth, λ is the wave length, T is the wave period, $k = 2\pi / \lambda$ is the wave number, and $C = \lambda / T$ is the so-called wave celerity. The wave period, wave length, and water depth are related by dispersion equation:

$$\lambda = T\sqrt{\frac{g}{k}\tan h(kh)}. \tag{5.3}$$

For shallow water ($h < 0.5\lambda$), it is shown that the wave length can be written as (Cornett 2008):

$$\lambda = \frac{gT^2}{2\pi}\left(\tan h\left[\left(\frac{4\pi^2 h}{gT^2}\right)^{3/4}\right]\right)^{2/3}. \tag{5.4}$$

For deep water ($h > 0.5\lambda$), $C = \lambda / T = 2C_g$, and $\lambda = gT^2 / 2\pi$; hence, the power of regular wave in deep water is written in the following form (Mørk et al. 2010):

$$P = \frac{1}{32\pi}\rho_{water}g^2 H^2 T. \tag{5.5}$$

Also, it is possible to show that the wave power per unit width transmitted by irregular waves can be approximated as:

$$P \approx \frac{\rho_{water}g}{16}H_S^2 C_g(T_e, h), \tag{5.6}$$

where H_S is the significant wave height, T_e is the so-called energy period and $C_g(T_e, h)$ is the group velocity. In deep water, the wave power can be rewritten in the following form:

$$P \approx \frac{\rho_{water}g^2}{64\pi}T_e H_S^2. \tag{5.7}$$

The energy period is rarely specified and must be estimated from other variables. For example, it is possible to relate the energy period and the wave peak period (T_p) as:

$$T_e = \chi T_p \tag{5.8}$$

The coefficient (χ) depends on the shape of the wave spectrum, e.g., for the Pierson–Moskowitz (PM) spectrum, $\chi = 0.86$ and for the Joint North Sea Wave Project (JONSWAP) spectrum with a peak factor of 3.3, $\chi = 0.9$. In some studies, the energy period is assumed to be equal to the wave peak period (Multon 2012):

$$P \approx \frac{\rho_{water}g^2}{64\pi}T_p H_S^2. \tag{5.9}$$

Table 5.1 Global and regional theoretical wave power resource (in GW). The net power (excluding areas where $P < 5$ kW/m and potentially ice-covered ones, Mørk et al. 2010)

REGION	P net
Europe (N and W)	286
Baltic Sea	1
European Russia	3
Mediterranean	37
North Atlantic Archipelagos	111
North America (E)	35
North America (W)	207
Greenland	3
Central America	171
South America (E)	202
South America (W)	324
North Africa	40
West and Middle Africa	77
Africa (S)	178
Africa (E)	127
Asia (E)	157
Asia (SE) and Melanesia	283
Asia (W and S) Asiatic	84
Russia	23
Australia and New Zealand	574
Polynesia	63
Total	*2985*

An area with a yearly average of energy more than 15 kW/m can be a possible site for the utilization of wave energy. The western seaboard of Europe (facing the Atlantic), the western seaboards of North and South America, southern Africa, Australia, and New Zealand are the best wave energy sources (Cornett 2008). Table 5.1 presents the global and regional theoretical wave power resource in GW. The net power, excluding areas where $P < 5$ kW/m and potentially ice-covered ones, is listed for different regions (Mørk et al. 2010). The global gross theoretical resource is found to be about 3.7 TW and the net resource is found to be about 3 TW (Mørk et al. 2010).

When waves get close to the shore, they lose some part of their energy due to effects of friction with the sea bottom. This phenomenon is discussed in detail in this book (refer to Chap. 8 "Wave and Wind Theories"). Hence, offshore waves in deeper water have more energy. This explains why it is better to put certain types of WECs offshore. However, there are some concepts that are designed and feasible for near-shore.

Most of the WECs are designed to work with resonances and have natural periods close to wave periods. Hence, they produce more power in regular wave patterns compared to irregular waves (keeping in mind that the wind-generated waves are basically irregular). Also, long-crested waves are better for power production than short-crested waves. Still, to calculate the possible energy in waves, long-crested regular waves can be implemented for start.

5.3   Wave Energy Converter Concepts

As mentioned earlier, hundreds of WECs have been introduced. Even though, there are just few concepts working in order and are feasible considering economics. The WECs can be classified based on the location and water depth where they are designed to function, e.g., shoreline, near-shore, and offshore. In practice, both kinetic and potential energy of the wave can be harnessed. Hence, based on how to absorb the energy, one may classify the WECs in three main groups (see Fig. 5.2):

1. Overtopping devices
2. Oscillating water columns (OWCs)
3. Wave-activated bodies

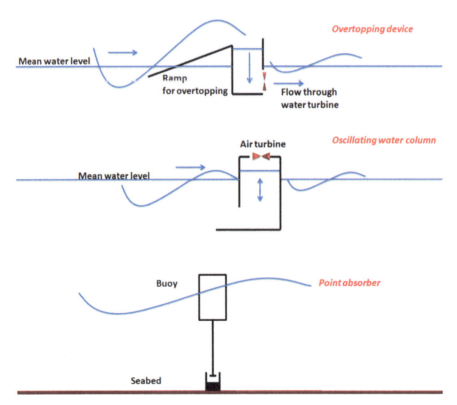

Fig. 5.2 Main categories of wave energy converters, overtopping device, oscillating water column and point absorber

5.4 Overtopping Energy Devices

The overtopping device applies an ascending ramp and the water spills over at a certain threshold. The increased potential energy can then be used. The water is collected and then goes through hydro turbines to generate electricity. Wave Dragon is a known example in this category. In the following, we discuss the overtopping concept while keeping Wave Dragon as an example in the entire section; this helps to understand the concept more easily.

Wave Dragon combines offshore technology and hydropower plants technology to produce electricity from waves using hydro turbines mounted in an advanced catenary-moored floating structure. The coming waves are guided over a ramp to be elevated above the sea level in a reservoir. The kinematic energy of waves is changed to potential energy in this way. In the next step, the water goes through hydro turbines and the electricity is produced. The low-head hydro turbine is the PTO system in this case. The stored water in reservoir should go out through hydro turbines between two coming waves. The reservoir has approximately 8000 m³ capacity. Main components of a Wave Dragon are (wavedragon.net 2005):

- Main body with a doubly curved ramp; a reinforced concrete and/or steel construction
- Two wave reflectors in steel and/or reinforced concrete
- Mooring system
- Propeller turbines with permanent magnet generators

Hydro turbines used in Wave Dragon are working separately (can be shutdown and started, independently) to make the power production smooth depending on the sea conditions. Wave Dragon uses traditional hydro propeller turbines with fixed gate vanes, which is a mature and well-proven technology that has been used in hydropower plants for more than 80 years (wavedragon.net 2005). A special, small-sized and low-headed turbine has been developed for Wave Dragon (e.g., a modified Kaplan turbine); refer to "Turbine Development for the Wave Dragon Wave Energy Converter" (Knapp et al. 2003).

Permanent magnet generator (PMG) is used for Wave Dragon. The rotation of the hydro turbines is transformed to electricity via a permanent magnet generator on each turbine (Sorensen 2004).

Apparently, turbines are the only moving parts in Wave Dragon. This is an advantage as moving parts are highly subjected to fatigue, and designing for survival load cases is a challenge as well. This means less-moving parts may lead to more cost-effective design. This has the advantages of reduction of maintenance costs and minimizing the harming effects of possible floating objects in the ocean (like debris). Still, the design of such system is comprehensive and needs considering optimized overtopping, stability of system, mooring system, structural components, and reduction in costs (capital, running and maintenance costs).

Wave dragon is a design for moderate water depth, e.g., 40 m. This helps harnessing of wave energy before they lose their energy due to approaching shoreline and interning shallow water areas. Wave Dragon is catenary moored using slack mooring lines. The ideal case is to have the system stabilized with less dynamics as much as possible (Kofoed et al. 2004).

Most of the marine structures, including ships and offshore platforms are designed in a way that the slamming, green water, wave on decks, and overtopping do not happen as much as possible. However, overtopping devices are designed to guide the waves in an efficient way with minimum loss of energy to overtop. For example, Wave Dragon has doubly curved ramp and reflectors in order to maximize the overtopping (wavedragon.net 2005). The ramp acts like a beach; the waves' kinetic energy converts rapidly to potential energy (wave changes its shape) as it goes up the short and steep ramp. Wave Dragon ramp has an elliptical shape to optimize this process. Numerical and experimental studies have proven positive effects of this elliptical shape on the power production.

Reflectors (see Fig. 5.3) are implemented in Wave Dragon to collect more waves and increase the power production. When waves reach the reflectors, they elevate and reflect toward the ramp increasing the amount of overtopping water, see Patent WO1996000848A1 (Friis-Madsen 2001).

Wave height changes over time; hence, the efficiency of overtopping devices is affected by wave height. Adjustment of draft of the overtopping device can help to maximize the power production. Wave Dragon uses pressurized air system to buoyant in correct level.

Fig. 5.3 Wave Dragon WEC seen from its reflector. (Courtesy of Erik Friis-Madsen (Friis-Madsen 2010). This file is licensed under the creative commons attribution 3.0 unported license)

Numerical simulations and model testing are the important phases of development of WECs. For overtopping devices, the numerical models estimating the overtopping of waves should be verified. Experiments in hydrodynamic laboratories have been carried out to fulfill this purpose (Parmeggiani et al. 2013).

Still, real-sea testing is needed to study the performance and integrity of these complicated systems. Overtopping WEC devices are comprehensive with respect to wave height, the geometry of the ramp and wave reflectors, the floating height of the platform, the amount of water overtopping, and the storing capacity of the reservoir.

Several parameters affect the design and performance of an overtopping device. In a separate chapter, design aspects of marine structures are discussed. Below, some important parameters are mentioned to have an idea about the main parameters that can affect WECs, especially, the overtopping marine energy devices.

1. Overtopping: The amount of water going to the reservoir is highly important and affects the power production. Free-board (adjustable in Wave Dragon), sea condition (actual wave height) and physical dimensions of the converter, such as ramp and reflector, influence the overtopping.
2. Outlet: The outlet means the amount of water passing turbines and representing power production. Size of reservoir, turbine design, and turbine on/off strategy affect the outlet.
3. Mooring system: Offshore overtopping devices are moored. For instance, Wave Dragon has catenary mooring lines. The orientation of the system relative to wave-dominant-direction can be fixed or free.
4. Size of the energy converter.
5. Metocean: The amount of energy in wave front (kW/m) and distribution of wave heights (short-crested waves and wave spectrum) have considerable effects on design.
6. Availability, reliability, maintainability, serviceability, redundancy, and maintenance are general topics for all kinds of marine structures, including overtopping WECs.

As an example, Wave Dragon was mostly constructed using standard components to reduce the maintenance cost. The strategy is to replace the turbine units with a schedule to increase the availability and reduce the costs (one may argue that it is not necessary to replace the components before they fail. The author is not supporting a concept or a specific idea; the main point is to discuss different methods and data regarding the design of marine renewable systems).

Another aspect of technology for WECs is the consideration of design for extreme environmental conditions. Overtopping structures are subjected to extreme loads in harsh conditions same as the other types. Adjusting the freeboard and designing the shape of the structure to minimize the loads are necessary when considering the high waves. The coupling of structure and the mooring system is also important to design a proper concept regarding both fatigue and extreme loads.

Table 5.2 Design parameters for different sites. (wavedragon.net 2005)

Wave Dragon key figures	Nissum Bredning 0.4 kW/m	24 kW/m	36 kW/m	48 kW/m
Weight, a combination of reinforced concrete, ballast, and steel	237 t	22,000 t	33,000 t	54,000 t
Total width and length	58 × 33 m	260 × 150 m	300 × 170 m	390 × 220 m
Wave reflector length	28 m	126 m	145 m	190 m
Height	3.6 m	16 m	17.5 m	19 m
Reservoir	55 m^3	5,000 m^3	8,000 m^3	14,000 m^3
Number of low-head Kaplan turbines	7	16	16–20	16–24
Generator's capacity for each turbine[a]	2.3 kW	250 kW	350–440 kW	460–700 kW
Rated power for a unit	20 kW	4 MW	7 MW	11 MW
Annual power production/unit	–	12 GWh/y	20 GWh/y	35 GWh/y
Water depth	6 m	>20 m	>25 m	>30 m

[a] Permanent magnet generator (PMG)

Beside the extreme loads and responses, it is necessary to study the possible fouling, which may stop moving parts and influence the structure (this includes debris, such as fishing nets, plastics, etc.).

The testing of the Wave Dragon model was done from 1998 to 2001 at Aalborg University. In 2003, a 20 -kW-rated WEC placed in Denmark (Nissum Bredning) that was the first grid connected offshore WEC (Kofoed et al. 2004). This prototype was fully equipped with hydro turbines and automatic control systems, and it was instrumented to monitor power production, wave climate, forces in mooring lines, stresses in the structure, and movements of the Wave Dragon.

The physical dimension of a Wave Dragon unit is optimized considering the wave climate at the offshore site. The width of the main body, length of the wave reflectors, weight, number, and size of turbines should be customized based on the offshore site data. The Wave Dragon tested in Nissum Bredning was constructed to match a moderate wave climate with approximately 0.4 kW/m energy resource. Table 5.2 presents main design parameters of device for different sites.considering their wave energy resources (wavedragon.net 2005).

The energy flux for a simplified long-crested regular wave depends on the depths of the water. Therefore, it is crucial how deep into the water the reflectors and the ramp reach. In shallow waters, the ramp must reach the sea bottom, while in deeper water, it must reach as far down as the wave itself. At 25–50-m water depth, the ramp reaches 10–13 m down (lorc.dk 2011a).

Fig. 5.4 Wavegen limpet wave energy turbine, in Scotland. (Pegrum 2006; the picture is modified by the author to include texts needed for clarification of device functions. © Copyright Claire Pegrum and licensed for reuse under this Creative Commons Licence)

5.5   Oscillating Water Column

An oscillating water column (OWC) works like a cylinder/piston. Wave energy makes water go up and down inside a chamber. By means of the fluctuating air pressure in the chamber, air turbines are rotating and electricity is produced (e.g., the land-installed marine-powered energy transformer (LIMPET) OWC5). Figure 5.4 shows a LIMPET in Scotland (courtesy of VOITH).

An OWC consists of a chamber in which the air is trapped, an opening below the water surface, and an air turbine. The water enters the chamber through a subsurface opening. Due to wave actions, the air inside the chamber goes up and down. Hence, the air compressed toward the air turbines when the wave is coming, and the air is sucked in from outside into the chamber when the waves are going. The OWCs are partially submerged, and the waves enter and exit from an entrance below the water surface. The air turbine produces electricity when the air passes through its blades. There are several ways to design such a concept; they may be fixed to the seabed, mounted on a floating structure, hanged from a shoreline structure, or built into harbor jetty. They are one of the most installed types of WECs. For 10 years, the LIMPET has been supplying power to the Scottish grid (see Fig. 5.4). Figure 5.5 shows the layout of a LIMPET.

The development of Wavegen LIMPET was continued by the installation of the Mutriku plant in 2011 (see Fig. 5.6). In 2011, the technology developed by Voith Hydro Wavegen became fully commercial. In the Basque seaport of Mutriku, between Bilbao and San Sebastian, a new facility has been integrated into the newly

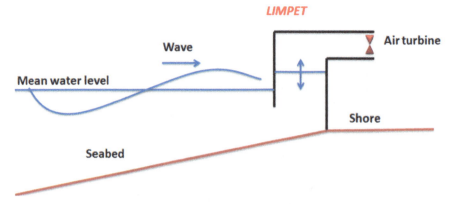

Fig. 5.5 Layout of a land-installed marine powered energy transformer (LIMPET)

constructed breakwater. Built for client Ente Vasco de la Energía (EVE), it consists of 16 Wells turbines, each 750 mm in diameter, rated at 18.5 kW and capable of generating almost 300 kW in total (voith.com 2011).

Fig. 5.6 Mutriku plant (300 kW). Author visit of the wave power plant in Spain. The picture shows a VOITH turbine with 18.5 kW installed power capacity

Fig. 5.7 Pelamis wave energy converter. (Courtesy of Guido Grassow (Grassow 2007). The file is licensed under creative commons attribution-share alike 3.0 unported license)

5.6   Point Absorber

Point absorber WECs are designed based on wave-induced motions of solid bodies. Usually, the moving part is tuned to have a natural period in the range of wave periods and hence, have resonance in operational conditions. For example, Pelamis (pelamiswave.com 2013) is a WEC consisting of several bodies moving relative to each other (see Fig. 5.7). The relative motion between moving bodies runs the generator. In some concepts, one body is moving relative to a moving reference point (Fig. 5.8) like Wavebob or relative to a fixed point, e.g., the PowerBuoy6.

The point absorber structures may be fully or partially submerged. For those WECs using floating buoys, the point absorber device is relatively small compared to a typical wave length and can absorb energy in all directions. A wide range of wind- generated waves produce electricity in these floating buoys and their foundation can be sea-bottom-mounted or floating structures. The PTO system may be a closed hydraulic system or a linear inductor (electric) depending to the design. Nonlinear PTO system is possible as well (Bailey 2009).

The point absorbers usually use translational motions (surge, sway and heave) to produce electricity. They are one of the main and early concepts among WECs since this industry appeared. Consider a floating buoy which has a generator at bottom of the sea (refer to Fig. 5.2); all translational motions, surge, sway, and heave, make a source for power generation in the PTO system. In general, the larger the buoy, the more energy can be absorbed as the exposed area to wave increases. However, if the buoy becomes too large, the diffraction forces are increasing and they dominate the

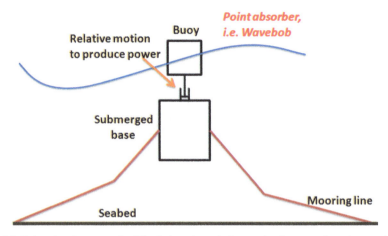

Fig. 5.8 Point absorber example, Wavebob, using relative motion between the submerged base and the floating buoy

loading and responses. Thus, the buoy will not work anymore as a point absorber (Engstrom 2011).

One of the most important parameters governing the design of WECs is the amount of produced electricity, and it is not just the amount of absorbed energy by the buoy which governs the design. Hence, it is needed to configure the entire system considering the dimensions, the mass and inertia of the system, and the PTO performance. This is usually done using an optimization algorithm for the selected offshore site and given metocean data. For example, an array of WECs can be more effective compared to a big size unit, for more information regarding the array of WECs (Ricci et al. 2007). Wavestar is an example of such an array in which the buoys are closely spaced (wavestarenergy.com 2011). The spacing ratio and dimension of buoys need precise optimization based on dominating wave periods. The dominant frequency of the waves in each sea state, e.g., the peak frequency, is a starting point of such studies.

Controlling the oscillation of WECs has a significant role on the power production. Both continuous and discrete control strategies are proposed. The continuous control strategy is possible if the information of incident waves and/or WEC's oscillations is available from measurements. The transfer functions are used to obtain the excitation forces. Convolution for excitation forces is done over a time interval (of length extending a few seconds into the past as well as into the future). The theoretically maximum converted power cannot be achieved by optimum controller, but only approached, because the prediction of physical variables, some seconds into the future, is imperfect. Correction of the waves is needed considering the radiated wave when measuring the incident waves. Otherwise, the error is significant in cases where a substantial fraction of the incident wave power is to be absorbed by the converter. With the known (and predicted) excitation force as input, the controller has to provide, as output, the optimum oscillating velocity (Falnes 1993).

Fig. 5.9 Two-body point
absorber, Lysekil project
wave energy converter. (ELE-
KTRICITETSLÄRA 2013)

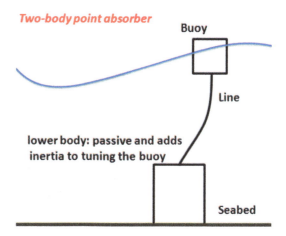

In discrete control, the oscillations are controlled just for a small number of times during a wave period. Latching was proposed for oscillating point absorbers (see, e.g., Budal 1978). When the velocity of the buoy is zero, for example, at the turning points, the body is locked. Afterward, when the velocity of the body is in phase with the predicted excitation force of the incident wave, the controller releases the buoy. In this way, latching helps to increase the PTO from the waves. The non-casual information of the excitation forces is needed. Latching helps to increase the power production and at the same time make a smoother output power (Falnes 2002).

Latching needs a breaking mechanism as well as enhanced controller. To avoid expensive and vulnerable control systems and breaking mechanisms, one way of shifting the response frequency of the system is increasing the inertia of the moving parts. This can be done using a point absorber system with two bodies (see Fig. 5.9). One body acting as a surface buoy, extracting energy, while the lower body is passive and adds the desired inertia tuning the point absorber to resonance (see, e.g., Engström et al. 2009). Both numerical simulations and the experimental setup have shown promising results for the two-body system with a 60 % absorption in irregular waves (Leijon 2008).

5.7   Wave Energy Converter Projects

As already mentioned, several 100 patents relating to the harvesting of wave energy have been introduced worldwide (lorc.dk 2011b). However, just a small number of these concepts have reached a maturity and have been installed; and, even fewer have produced energy for the grid. In this section, some of the mature projects are presented.

Fig. 5.10 Wavestar. (Courtesy of Sebastian Nils Swiatecki (Swiatecki 2011). This file is made available under the Creative Commons CC0 1.0 Universal Public Domain Dedication)

5.7.1   Wavestar (Wave Star Energy)

Wavestar (Fig. 5.10) is a set of point absorbers on a unit based on the definition given in this book. However, in some literatures, this type is also called "attenuator" (see, e.g., Drew et al. 2009). Attenuators lie parallel to the dominant wave direction and "ride" the waves. The Wave Star Energy considers it a multi point absorber system (Kramer 2006).

Niels and Keld Hansen invented the concept in 2000. They tried to make a regular energy output from swell and wind-generated waves. This was achieved with a row of half-submerged buoys riding (rise and fall) as the wave passes. This forms the iconic part of Wavestar's design and allows energy to be continually produced despite waves being periodic (wavestarenergy.com 2011).

Some of the characteristics of the Wavestar concept are listed below (Kramer 2006).

- Using standard components, e.g., from offshore wind turbine technology
- Installed on a pile structure on the seabed—in 7–30 m of water depth
- It can be shut down in harsh conditions to protect the structure, for example, if waves exceed 8 m

The power performance of the Wavestar concept for 40 floats is listed in Table 5.3. In Table 5.4, the project developments during the past years are listed.

Table 5.3 Wavestar power performance with 40 floats. (Kramer 2006)

Scale	Depth (m)	Length (m)	Floaters diameter (m)	Significant Wave Height (Hs) (m)	Power (kW)
1:10	2	24	1	0.5	1.8
1:2	10	120	5	2.5	500
1:1	20	240	10	5	6000

Table 5.4 Wavestar project development during the past years

Year	Facts/events
2000	Wavestar was invented
2003	Company bought the right
2004	1:40 scale model tested in Alborg University
2005	1:10 scale grid connected built
2006	1:10 scale grid connected installed offshore
2009	1:2 scale with 600 kW installed offshore and connected to grid since 2010

Wavestar consists of two rows of point absorbers. Each row has 20 floats which are moving up and down, separately, when the wave passes. The rows are parallel with the dominant direction of waves. Each float is connected to the main hydraulic motor. The hydraulic motor runs the generator to make electricity. When the float is going up, it pushes the hydraulic fluid toward the motor, and when it comes down, it sucks the hydraulic liquid from the outlet of the motor (this is a closed hydraulic loop). The floats are relatively close to each other and they smoothly follow the passing wave Fig. 5.11.

5.7.2 Pelamis (Pelamis Wave Power)

Pelamis (Fig. 5.7) is a set of point absorbers connected to each other and it produces electricity by using relative motion between adjacent parts. Pelamis is like a snake and it is a floating device consisting of a few floating parts. For example, one of the prototypes tested was 120 m long and 3.5 m in diameter. It is comprised of four tube sections linked by three, shorter, power conversion modules.

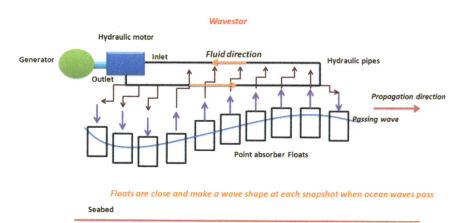

Fig. 5.11 Wavestar schematic layout for energy production; floats are set relatively close to each other. (Courtesy of wavestarenergy.com 2011)

Table 5.5 Pelamis development during the past years

1998	Start of the project by model testing
2004–2007	Full-scale prototype machine was tested at the European Marine Energy Centre
2006	Prototype was upgraded
2007	Installation of upgraded prototype on the site of Engineered Manufacturing and Equipment Company (EMEC)
2008	Installation of three devices (total capacity of 2.25 MW) in Portugal at Aguçadoura
Now	Pelamis P2 design was sold to utility customers E.ON and ScottishPower Renewables. P2 machines are currently being tested for a number of commercial scale projects

Pelamis is moored by a catenary mooring line to seabed. Like Wavestar, it is considered an "attenuator" WEC device. Pelamis is developed by Ocean Power Delivery Ltd, which is now known as Pelamis Wave Power. Pelamis operates in water depths greater than 50 m and is typically installed 2–10 km from the coastline. The rated power is 750 kW with a target capacity factor of 25–40%, depending on the offshore site and environmental conditions. The Pelamis prototype was the world's first commercial scale WEC to generate electricity to a national grid from offshore waves (pelamiswave.com 2013a). Pelamis has inherent weather warning and sets itself parallel facing with coming waves. Hence, the device is always set facing (parallel with) the dominant direction of the waves Table 5.5.

Pelamis P2 has five tubular sections linked by joints that allow movement in two directions. The structure floats semi-submerged on the surface of the water and inherently faces into the direction of the coming waves. The length of the system is in the order of operational wave length while the cross section is small compared to waves.

As waves pass the length, the sections bend in the water (see Fig. 5.12). The relative movement in joints is converted into electricity via hydraulic PTO systems. Each joint has its own separate PTO system. The PTO system is driven by hydraulic cylinders at the joints that resist the wave-induced motion and pump fluid

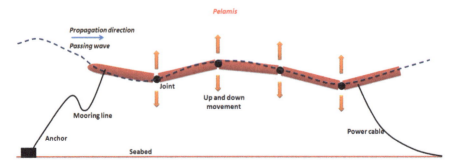

Fig. 5.12 Pelamis schematic layout for energy production. (pelamiswave.com 2013b)

into high-pressure accumulators, which allow power generation to be smooth and continuous. The electrical power is transmitted to coastline using standard subsea cables and equipment.

5.7.3   Wave Dragon (Wave Dragon A/S)

Wave Dragon is installed at Nissum Bredning in Denmark. It is of the device type overtopping terminator. In Sect. 5.4, this project has been thoroughly discussed.

5.7.4   OE Buoy (Ocean Energy Ltd.)

OE Buoy (see Fig. 5.13 which shows the EMEC scale test) is a floating OWC device moored to seabed using mooring lines. The concept is designed by Ocean Energy Ltd. that is a commercial company developing wave energy technology. The device has been tested for more than 2 years of sea trials in Atlantic waves

Fig. 5.13 Test buoy for EMEC wave scale test site, scapa flow. (Curtis 2012, © Copyright Andrew Curtis and licensed for reuse under this Creative Commons Licence)

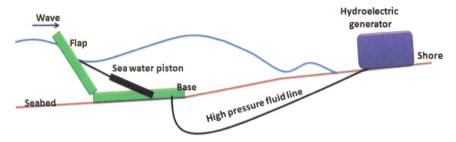

Fig. 5.14 Oyster wave energy converter, schematic layout of the device. (aquamarinepower.com 2013)

(oceanenergy.ie 2013). OE Buoy is in its development process. After successful test of 1/50 scale of the device in University College Cork, Ireland and 1/15 scale in Ecole Central de Nantes, France, a 1/4 scale (28 t) model was launched for stability sea trials in 2006. The OE buoy has been in operation more than 24,000 hours in the Atlantic Ocean. Wells and impulse turbines, which both of them are self-rectifying types, are used in OE buoy to compare the performance of such turbines for OWC. The advantage of self-rectifying turbines is that they can take off power in both directions of the air flow. Lavelle and Kofoed showed that the estimated yearly power production of the OE Buoy at the test location was 12.7 MW·hour/year, and the peak efficiency was approximately 17 % (Lavelle and Kofoed 2011).

5.7.5   Oyster (Aquamarine Power)

Oyster is a nearshore point absorber WEC. Aquamarine Power's Oyster wave power technology captures energy in nearshore waves and converts it to electricity. Oyster is a wave-powered pump which pushes high-pressure water to drive an onshore hydroelectric turbine. The Oyster wave power device is a buoyant, hinged flap, which is attached to the seabed at the depths between 10 and 15 m, around half a kilometer from the shore (aquamarinepower.com 2013).

Oyster has a hinged flap, which is almost entirely underwater. The flap pitches backward and forward due to the action of the nearshore waves. This movement of the flap drives two hydraulic pistons. The pistons push high-pressure water to onshore using a subsea pipeline to drive a conventional hydroelectric turbine (see Fig. 5.14). When an array of WECs are installed, it is possible to use a main subsea pipeline, which is connecting multiple Oyster wave energy devices to a single onshore plant. Aquamarine has an ultimate plan to install wave farms of several 100 Oyster energy devices generating hundreds of megawatts of electrical power (Collier et al. 2008).

Fig. 5.15 WaveRoller wave energy converter. (aw-energy.com 2013)

5.7.6 WaveRoller (AW Energy)

WaveRoller (refer to Fig. 5.15) designed by AW Energy is a point absorber WEC device. WaveRoller is a nearshore energy device that converts ocean waves to electrical power. It is usually installed with an approximate distance of 0.3–2 km from the shore at water depths of 8–20 m. It consists of a fully submerged flap that is anchored to the sea bottom. A single WaveRoller unit (one panel) has rated power between 0.5 and 1 MW. The capacity factor of the system varies between 0.25 and 0.5 depending on environmental conditions at the offshore site (aw-energy.com 2013).

The concept works based on a physical phenomenon so-called "wave surge" that occurs when waves approach the shore line (Westhuysen 2012). Waves in deep water (deeper than half the length of the wave) are basically water particles moving in a circular motion. However, in nearshore sites, the waves start "shoaling" since some of the water particles come into contact with the seabed. Due to the interaction with the seabed, the wave's motion gets horizontally elliptic shape rather than circular in deep water. The water particles are flattened and stretched in such small depth. Also, horizontal movement of water particles in shallow water gets amplified and a strong surge zone is created. This amplifies the back and forth movement of water in nearshore, which can be deployed to drive the WaveRoller flap and creates energy. The back and forth movement of water driven by wave surge puts the composite panel into motion.

The WaveRoller is designed for water depth of 8–20 m in which the wave surge power is high. This maximizes the power production; the flap is extended from the sea bottom to below the water surface without breaking the surface to ensure limiting the load and increase the structural integrity.

The back and forth movement of the flap runs the hydraulic piston, which pumps the hydraulic fluid to a closed circuit. All the elements of this hydraulic circuit are inside a hermetic structure inside the WaveRoller device and hence are not exposed to sea water. So, there is no risk of leakage into the ocean. The high-pressure fluids are fed into a hydraulic motor that drives a generator. The electrical power is then connected to the electric grid through a subsea cable. Arrays of WaveRoller, e.g., tens of devices can be installed in a single site to reduce the cost by sharing the infrastructure among the machines. Each WaveRoller is equipped with an on-board electricity generator. So, the output from many devices can be combined via electricity cables and a substation. This reduces the cost of produced electricity from the farm.

5.7.7 LIMPET (Voith Hydro Wavegen Ltd.)

LIMPET is a device of the OWC type (refer to Sect. 5.5 for more information about this project).

5.7.8 OceanLinx (Oceanlinx)

OceanLinx is an OWC WEC made of concrete. Oceanlinx has demonstrated three large-size test platforms in the ocean over the past 16 years. Oceanlinx was the first company to achieve the full grid connection of a test platform in Australia in 2010.

The greenWAVE unit was already built in October 2013, and it is due to be installed in Port MacDonnell, Australia. It is the world's first 1-MW single unit WEC (oceanlinx.com 2013). The structure is constructed and will be launched into the water from TechPort in Adelaide. The US$ 7 million project will be the first of its kind to be commissioned in the world. The unit will be grid connected and rigorously tested in 2014. The greenWAVE is a single OWC designed for shallow water applications. It is made of simple-packed prefabricated reinforced concrete. The structure is bottom fixed and sits under its own weight (3000 t) on the seafloor without the need of seabed preparation. The unit is built environmentally friendly without anchors, mooring, or attachment to the seabed. Also, there are no moving parts under water. The top of the structure is used for housing the air turbine (so-called "airwave" turbine) and electrical control systems (oceanlinx.com 2013).

5.7.9 CETO (Carnegie Wave Energy Limited)

CETO is a submerged point absorber WEC. CETO wave power converter is fully-submerged and produces high-pressure water from the waves. By delivering high-pressure water ashore, either electrical power (e.g., hydroelectricity) or freshwater (utilizing standard reverse osmosis desalination technology) can be produced (carnegiewave.com 2013).

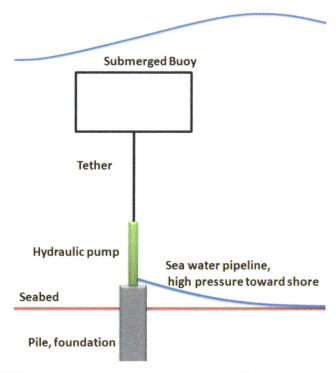

Fig. 5.16 CETO wave energy converter. (carnegiewave.com 2013)

CETO 5 has 240 kW rated power capacity, which is increased three times compared to the previous generation, CETO 3. The CETO 3 unit was tested at the Garden Island site in 2011. The 10-m-diameter CETO 4 unit is being deployed by EDF and DCNS off Reunion Island. The diameter of the buoyant actuator in CETO 5 has the most significant influence on power output and has been increased to 11 m from the 7 m diameter (CETO 3).

The initial development of CETO began in 1999. Work on the design of the CETO technology platform commenced in 2003 with the construction of the first prototype unit. The CETO 1 prototype proved the concept of generating zero-emission power and freshwater from the ocean waves in 2006. Between 2006 and 2008, CETO 2 prototypes were developed and tested in the waters of Fremantle.

The concept consists of the following items (see Fig. 5.16):

1. Buoyant actuator: This is a buoy submerged few meters below the water surface. Due to wave action, the buoy moves up and down. The buoy is symmetric and hence wave direction is not affecting the performance of the concept.
2. Tether: The tether is connected to buoy in one end and to the pump at the other end. The energy captured by the buoy is transferred to pump using the tether.

The design of the tether can be based on common practice in the offshore oil/gas industry. Tether flexibility limits transferring the unwanted loads to the system.

3. Pump: The pump is the PTO system in this concept. It converts the energy of waves to hydraulic pressure and pumps the sea water toward shore through pipelines.
4. Foundation: Drilled or grouted pile foundation is used. The foundation anchors the pump to the seabed.

5.7.10   Powerbuoy (Ocean Power Technologies)

Powerbuoy is a point absorber WEC with a floating-base moored to the seafloor. Ocean Power Technologies (OPT) Powerbuoy is designed to convert ocean wave energy into useable electrical power for grid-connected applications. The Powerbuoy can be deployed in arrays scalable to hundreds of megawatts (oceanpowertechnologies.com 2013). The company designed the following Powerbuoy devices:

• Autonomous PowerBuoy
• Mark 3 PowerBuoy
• Mark 4–The PowerTower

OPT is currently developing the PowerTower, a Mark 4 PowerBuoy, that is planned to drive a generator with a 2.4-MW peak rating. The US Department of Energy (DOE) and the UK Government are funding several stages of the Mark 4 development. The Mark 4 PowerTower is expected to improve on the customer value proposition of the Mark 3 PowerBuoy and expand the available market areas.

The first utility-scale Mark 3 PowerBuoy, fabricated in Scotland, was deployed in 2011 off the Eastern coast of Scotland for ocean trials. A second Mark 3 PowerBuoy is being fabricated in Portland, OR, and is planned for deployment in OR (oceanpowertechnologies.com 2013). The Mark 3 generates power with wave heights between 1 and 6 m. The device is typically configured in arrays of two to three rows to minimize the footprint.

Mark 3 PowerBuoy drives a generator with 866 kW rated power. The typical capacity factor of Mark 3 is 0.3–0.45 (this depends on location and environmental conditions). The PowerBuoy has fiber-optic communications and supervisory control and data acquisition (SCADA) systems.

In January 2011, OPT achieved the Lloyd's Register certification for the Mark 3 (see Fig. 5.17) for its intended use, as analyzed against international standards, and its survivability in severe wave conditions.

An array of Mark 3 consists of the PowerBuoys, an undersea substation pod (USP), and the transmission cables to shore. Up to ten Mark 3 PowerBuoys can be connected to each USP (oceanpowertechnologies.com 2013). The buoy moves due to wave actions, and the relative displacement (vertical motion) between the buoy

Fig. 5.17 PowerBuoy wave energy converter. (Duckworth 2011. © Copyright Sylvia Duckworth and licensed for reuse under this Creative Commons Licence)

Table 5.6 Mark 3 Power-buoy specification. (ocean-powertechnologies.com 2013)	Rated power	866 kW
	Overall length	43.5 m
	Height above waterline	11.5 m
	Float diameter	11 m
	Weight	180 t
	Design life	25 years
	Mooring	Three point
	Deployment	Tow out with standard tug
	Wave height	1.6 m
	Water depth	55 m (minimum)

and floating foundation (Spar) runs a mechanical system coupled to generators which produces alternating current (AC). The electricity is rectified and inverted into grid-compliant AC. The specification of Mark 3 Powerbuoy is listed in Table 5.6 (Fig. 5.18).

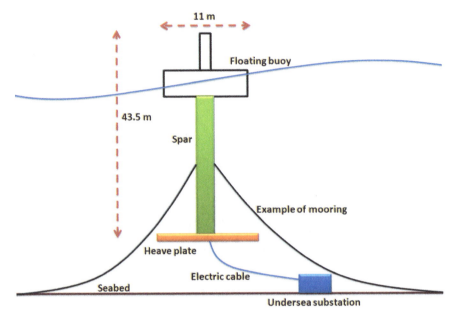

Fig. 5.18 Schematic layout of Mark 3 PowerBuoy wave energy converter. (oceanpowertechnologies.com 2013)

References

aquamarinepower.com. (2013). Technology: How Oyster wave power works. http://www.aqua-marinepower.com/technology/how-oyster-wave-power-works/.

aw-energy.com. (2013). WaveRoller concept. http://aw-energy.com/about-waveroller/waveroller-concept.

Bailey, H. (2009). *The effect of a nonlinear power take off on a wave energy converter*. Edinburgh: The University of Edinburgh.

Budal, K. (1978). Patent No. United States Patent 4203294. US.

carnegiewave.com. (2013). What is CETO. http://www.carnegiewave.com/index.php?url=/ceto/what-is-ceto.

Collier, D., Whittaker, T., & Crowley, M. (2008). *The construction of oyster—A nearshore surging wave energy converter*. International Conference on Ocean Energy, Brest, France.

Cornett, A. M. (2008). *A global wave energy resource assessment*. International offshore (ocean) and polar engineering conference, (ISOPE, Paper No. ISOPE-2008–579), Canada, Vancouver.

Curtis, A. (2012). Test buoy for EMEC wave scale test site, scapa flow. http://www.geograph.org.uk/photo/3101220.

Drew, B., Plummer, A. R., & Sahinkaya, M. N. (2009). A review of wave energy converter technology. *Proceedings of the IMechE Part A: Journal of Power and Energy, 223*, 887–902.

Duckworth, S. (2011). Ocean power technology pb150 wave-power generator. http://www.geograph.org.uk/photo/2669058.

ELEKTRICITETSLÄRA. (2013). Wave power project—lysekil. http://www.el.angstrom.uu.se/forskningsprojekt/WavePower/Lysekilsprojektet_E.html#Historia.

Engstrom, J. (2011). *Hydrodynamic modelling for a point absorbering wave energy converter*. Uppsala: Uppsala University.

Engström, J., Eriksson, M., Isberg, J., & Leijon, M. (2009). Wave energy converter with enhanced amplitude response at frequencies coinciding with Swedish west coast sea states by use of a supplementary submerged body. *Journal of Applied Physics, 106*(6), 064512–064515.

Falnes, J. (1993). Optimum control of oscillation of wave-energy converters. www.ntnu.no: http://folk.ntnu.no/falnes/web_arkiv/InstFysikk/optcontrl.pdf

Falnes, J. (2002). *Ocean waves and oscillating systems*. Cambridge: Cambridge University Press.

Falnes, J. (2007). A review of wave-energy extraction. *Journal of Marine Structures, 20*, 185–201.

Friis-Madsen, E. (2001). Patent No. EP 0767876 B1.

Friis-Madsen, E. (2010). Wavedragon. http://commons.wikimedia.org/wiki/File:WaveDragon.JPG.

Grassow, G. (2007). Pelamis Wellenkraftwerk Portugal 3. http://commons.wikimedia.org/wiki/File:Pelamis_Wellenkraftwerk_Portugal_3.JPG.

Knapp, W., Böhm, C., Keller, J., Rohne, W., Schilling, R., & Holmén, E. (2003). Turbine development for the Wave Dragon wave energy converter. (Proceedings, Hydro, Vol. I, Croatia, pp. 367 –376). http://citeseerx.ist.psu.edu/viewdoc/download?doi=10.1.1.197.9237&rep=rep1&type=pdf.

Kofoed, J. P., Frigaard, P., Friis-Madsen, E., & Sørensen, H. C. (2004). *Prototype testing of the wave energy converter wave dragon*. In A. A. M. Sayigh (Ed.), *World renewable energy congress VIII (WREC 2004)*. Denver: Elsevier Science.

Kramer, M. (2006). The wave energy converter WaveStar, a multi point absorber system. http://wavestarenergy.com/sites/default/files/Wave%20Star%20Bremerhaven%20Kramer.pdf.

Lavelle, J., & Kofoed, J. P. (2011). *Power production analysis of the oe buoy wec for the cores project*. Aalborg: Aalborg University, Department of Civil Engineering, Wave Energy Research Group.

Leijon, M., et al. (2008). Wave energy from the north sea: Experiences from the Lysekil research site. *Surveys in Geophysics, 29*, 221–240.

lorc.dk. (2011a). Energy in waves. http://www.lorc.dk/wave-energy/energy-in-waves.

McCormick, M. E. (2007). *Ocean wave energy conversion*. USA: Dover.

Mørk, G., Barstow, S., Kabuth, A., & Teresa Pontes, M. (2010). *Assessing the global wave energy potential*. 29th international conference on ocean, offshore mechanics and arctic engineering (Shanghai, China). OMAE2010–20473.

Muetze, A., & Vining J. G. (2006). *Ocean wave energy conversion—A survey*. University of Wisconsin-Madison, IEEE, USA.

Multon, B. (2012). *Marine renewable energy handbook*. UK: Wiley.

oceanlinx.com. (March 2013). New look greenWAVE. http://www.oceanlinx.com/.

oceanpowertechnologies.com. (2013). Utility scale systems. http://www.oceanpowertechnologies.com/products.html.

Parmeggiani, S., Kofoed, J. P., & Friis-Madsen, E. (2013). Experimental update of the overtopping model used for the wave dragon wave energy converter. *Energies, 6*(4), 1961–1992. doi:10.3390/en6041961.

Pegrum, C. (2006). Wave energy power plant, Islay. http://www.geograph.org.uk/photo/273216.

pelamiswave.com. (2013a). Pelamis wave power ltd. http://www.pelamiswave.com/.

pelamiswave.com. (2013b). pelamis-technolog. http://www.pelamiswave.com/pelamis-technology.

Ricci, P., Saulnier, J. B., & de O. Falcão, A. F. (2007). *Point-absorber arrays: A configuration study off the Portugese West-Coast*. Proceedings of the 7th European Wave and Tidal Energy Conference, Porto, Portugal.

seabased.com. (2013). Bringing wave energy to the world. http://www.seabased.com/en/.

Seed, M. (2012). Wavegen presentation. http://www.slideshare.net/bemadrid/wavegen-presentation.

Sorensen, H. C. (5–9 Sept 2004). *World's first offshore wave energy converter –wave dragon—connected to the grid*. 19th World Energy Congress, Sydney, Australia.

Straume, I. (2014). Wave energy concepts overview numbered. http://commons.wikimedia.org/wiki/File:Wave_energy_concepts_overview_numbered.png.

Swiatecki, S. N. (2011). SskWave Star. http://commons.wikimedia.org/wiki/File:SskWave_Star.
    jpg.
Tedd, J. (2007). *Testing, analysis and control of wave dragon, wave energy converter*. PHD Thesis,
    Aalborg University, Denmark.
voith.com. (2011). HyPower. http://voith.com/en/HyPower_20_final.pdf.
wavedragon.net. (2005). Wave dragon. http://www.wavedragon.net/.
wavestarenergy.com. (2011). http://wavestarenergy.com/concept. http://wavestarenergy.com/,
    http://wavestarenergy.com/sites/default/files/EWTEC2011_2011-09-06.pdf
Westhuysen, A. J. (25–27 June 2012). *Modeling nearshore wave processes*. ECMWF workshop
    on ocean waves.

Chapter 6
Combined Wave- and Wind-Power Devices

6.1 Introduction

In the previous chapters, the wind turbines and wave-energy converters (WECs) are separately explained. Land-based wind turbines and their key components as well as offshore wind turbines are discussed. Examples of existing projects for both fixed and floating wind turbines are given while the main supporting structures for the offshore wind industry are introduced. The WECs and their main categories are explained, and some examples of current projects in the world are mentioned. This chapter focuses on the hybrid marine platforms in which the wave- and wind-energy devices are combined to use the possible synergy effects and reduce the cost of electrical energy from offshore units while increasing the quality of the delivered power to the grid.

In the past decade, more offshore wind farms have been constructed. Hence, the possibility of integrating other marine renewables, such as WECs and ocean current turbines with offshore wind, are increased. The integration can present several advantages, such as better utilization of the ocean space and decreasing the associated costs, e.g. installation/maintenance costs relative to separate installations. There are possibilities to share substructures and infrastructures between the devices as well. Here, the focus is on integrating the wind- and wave-power devices to present hybrid concepts.

Both WECs and offshore wind turbines are subjected to similar challenges, such as harsh marine environmental conditions. However, the maturity of them is different. WECs came earlier to offshore business than offshore wind turbines. But, offshore wind technology had a good development especially in the past decade compared to wave energy. Offshore wind is a proven technology with 3.8 GW of installed capacity in Europe (Mocia et al. 2001). By the integration of wave to wind, there is a better chance for wave energy to reduce cost; taking advantage of the fact that the offshore wind energy industry is more mature.

The combination of wave and wind energy has two aspects. One is combining the power production of wave-energy devices and wind turbines in a farm, e.g. independent bottom-fixed wind turbines and WECs. The second format, which is

© Springer International Publishing Switzerland 2014
M. Karimirad, *Offshore Energy Structures*, DOI 10.1007/978-3-319-12175-8_6

the main focus of this Chapter, is combining the wave- and wind-energy devices in one unit so-called hybrid platform, e.g. bottom-fixed hybrid wind-WECs or floating hybrid wind-wave-energy converters. Both combination forms, (a) segregated and (b) hybrid, are discussed; however, the core is the hybrid platform, its feasibility, proposed concepts and advantages achieved by such combinations.

6.2   Combining Offshore Wind and Wave Energy, Why?

The main point in combining a wind turbine concept with a WEC concept is to decrease the finished cost of electricity in a long-term perspective. No matter from which side we start; from wave energy or from wind power. The goal is clear which is maturing the offshore renewable energy. The idea of adding a wave-energy device to a floating support structure of a wind turbine is increasing the stability, damping the structure motion by extracting the wave energy and, hence, increasing the produced wind power while taking the wave power, simultaneously.

In a good design (advanced combined concept), the wave-energy device acts like a damper which takes out the coming incident wave energy and therefore increase the stability of the support structure. This means less motion of the floating wind turbine and hence increased the power. On the other hand, the WEC produces electricity in addition to the wind turbine. There are several other advantages when combining wave- and wind-energy devices (if it is properly performed), such as sharing the electrical subsea cables, survey and monitoring costs as well as sharing the supporting structures, foundation, mooring and anchoring systems.

In some concepts, due to wave-power absorption, the water is getting calmer at the rear of the support structure, and therefore the boat landing is easier, which helps monitoring and maintenance to be done with a lower cost and less challenges. Looking at the other side of the coin, adding one or several wind turbines to a floating wave-energy device, increase both stability and power production. The support structure will be stronger and heavier for a given capital cost, which means more stable base for the wave energy take-off system. Although, since the wave-energy device needs a floating support structure in some cases, the cost of including wind turbines offshore in such a system is close to the expense of having the turbines onshore.

Poseidon is an example of hybrid platforms combining wave energy and wind power in one unit to use the synergy and increase the power production as it is mentioned above. Later in this chapter, more information is given for this concept. Figure 6.1 shows Poseidon floating power platform in operation with both WECs and three wind turbines, all integrated in one offshore unit.

In a wind farm, wind turbines should have a certain distance from each other, e.g. 5 times of the rotor diameter. This is due to wake effects and air turbulence. This means the MW-scale wind turbines are away from each other in order of half a kilometre. However, the wave plants can be set in a dense pattern between wind turbines. This results in even more power in less ocean space. By such a combination

Fig. 6.1 Poseidon, an example of hybrid platforms combining wave- and wind-energy devices. (Courtesy of Sam Churchill (Churchill 2011). Original content is released under the CC-BY-SA license)

of wave and wind farms, installation and maintenance costs are lower, since the farms can share the same:

- Geographic footprint
- Interconnection cabling
- Power switch
- Undersea cable to the grid
- Boats
- Transformer platform
- Technical team/Personnel
- Permissions cost

Several advantages can be achieved by combining the wave- and wind-energy devices either (a) integrated in one unit or (b) segregated close to each other. When the WECs and offshore wind turbines farms are built close to each other, they can share cost, infrastructure and services to reduce the cost and improve the electrical power quality. By combining the offshore wind and wave-energy devices in one unit, more benefits can be achieved e.g. improving the dynamic motions and consequently increasing the produced power. Some of the benefits achieved in a combination of wave- and wind-energy devices in hybrid platforms are listed below:

- Reduced weight by sharing the platform foundations
- Cost reduction for design/operation, sharing cost for:
  - Common infrastructure and equipment onshore
  - Offshore transmission infrastructure
  - Sharing grid connections
  - Permitting and project-development costs
  - Operating, monitoring and maintenance costs
- Increase renewable energy production per surface area of ocean
  - Optimization of the utilization of sea areas
- Utilizing simultaneously two sources of energy

- Reduce the hours of zero produced electricity
  - Significantly larger time windows for power delivery
- Increase the capacity value of the farm
- Reduce the variability of the produced electricity
- Smooth power output
  - Improvement of efficiency
  - Better quality of the power
- Reduce the forecast error of the power output
- Less required transmission capacity

The idea of combining the power generation with wind and wave is not new; for example, refer to Lakkoju (1996). Lakkoju discussed the advantages of "Combined Power Generation with Wind and Ocean Waves" and tried to statistically demonstrate some advantages of combined wind and ocean wave power generation. One of the major advantages of combined wind and wave-power generation is to improve the probability of continuous power supply (it minimizes the interruptions and compensates power fluctuations of one with another). This minimizes one of the major criticisms for renewable sources of energy. Lakkoju showed promising results and indicated that the combined power generation improves the probability of continuous power supply.

Fusco et al. have shown how the combination of wave and wind causes variability reduction for an Irish case study (Fusco et al. 2010). Analysis of the raw wind and wave resources in some locations in Ireland shows that they are low correlated. Hence, the integration of wind and waves in combined forms allows the achievement of a more reliable, less variable and more predictable electrical power production (Fusco et al. 2010).

Stoutenburg and Jacobson studied the effect of resource diversity, such as combining offshore wind and wave energy for reducing the impact of variability on the electric power system and facilitating higher presentations of renewable. The power output profiles of a 100 % wind farm, a 100 % wave farm, and combined farms with wind and wave were shown to be significantly different from each other; and, this difference resulted in a lower transmission capacity requirement for the combined farms (Stoutenburg and Jacobson 2010).

Chozas et al. showed that the "balancing cost" is reduced for a combined wave and wind system. There is usually a cost associated to the integration of non-fully predictable renewable, such as wave or wind in electricity markets. This cost, named "balancing cost", covers the difference between the bid to the day-ahead electricity market and the actual power produced. They showed that when wave converters are combined the balancing costs keep low, 45 % lower than for wind turbines. Finally, a diversified scenario of wind and wave technologies brings balancing costs 35–45 % down compared to the only-wind scenario (Chozas et al. 2012).

Chozas studied the combination of Wavestar with a wind turbine. The results showed that the best combination is 50 % wind plus 50 % wave energy. This resulted in barely any zero-power production and less fluctuation of electricity (Chozas 2012). The combined wave- and wind-energy devices in one unit as a hybrid marine energy platform is being more accepted. In the recent years, more research has been carried out to study the feasibility of such combination and highlighting the positive effects gained by such combination of energy production in one unit.

6.3   Poseidon: An Example of Combining Wave and Wind Devices

In this section, Poseidon floating power plant is studied as an example of operating projects combining the wave-energy and wind-power devices in one unit. Floating Power Plant (floatingpowerplant.com 2012) has constructed a 37 m model of the Poseidon concept for a full offshore test at Vindeby offshore site which is located off the coast of Lolland in Denmark. The demonstration project so-called Poseidon 37 is 37 m wide, 25 m long, 6 m high (to deck, not including the turbines) and weighs approximately 320 t. The commercial width range of Poseidon is between 80 and 150 m depending on the environmental conditions. Poseidon 37 was launched in Nakskov Harbour and installed at the offshore site in 2008.

The Poseidon concept was established back in 1980 (floatingpowerplant.com 2012). In 1996, the development process was speeded up, and the concept has during the last decade undergone tests in scale models of the following sizes before sea trial of the Poseidon 37:

- 2.4 m wave front, system test (scale ~ 1/30 of a 80 m unit)
- 15 m wave front, floater test (scale ~ 1/5 of a 80 m unit)
- 8.4 m wave front, system test (scale ~ 1/10 of a 80 m unit)

2.4 m Wave Front, System Test   In 1998, the first concept test was performed at Aalborg University. The test was performed without wind turbines. The aim of the test was to verify the durability and sustainability of the concept. The results showed potential for a new competitive wave power take-off system (PTO system).

15 m Wave Front, Floater Test   Between 1999 and 2000, the floaters for the concept were tested in a wave flume in a system scale of 15 m wave front (e.g. the tested floaters had a 0.75 m wave front). The goal of the test was optimizing the design of the floaters, shape and ballasting, with respect to energy utilization.

8.4 m Wave Front, System Test   Between 2001 and 2002, the 8.4 m model with simulated wind turbines was tested. The results documented a utilization rate of 0.35 from wave energy to electricity. The concept was also documented as a floating foundation for offshore wind turbine. Furthermore, it was found that the use of wind turbines increased the utilization rate of the wave power plant. This is due to less movement of the platform (floatingpowerplant.com 2012).

Table 6.1 Characteristics of the Poseidon 37. (floating-powerplant.com 2012)

Parameter	Value
Total rated power	177 kW
Wave-energy-rated power	140 kW
Wind turbines-rated power	3×11 kW
Turbines	Three downwind
Floater	Semi-submerged, ship-shaped
Mooring	Turret
Water depth	More than 40 m
Weight	320 t
Ballast	40 t
Width	37 m
Length	25 m
Draft	4.5 m
Height of floater (to deck)	6 m

Poseidon Structure The structure of Poseidon consists of three parts. The front part contains the turret mooring, the middle part carries the WECs and two wind turbines; and, the rear part carries one wind turbine. The landing to the unit is done through the rear part as the wave energy is taken from the middle part by WECs.

The middle part can be disconnected from the front, and the energy device can be transported to quay side without interfering with the mooring. The front and rear sections ensure that the Floating Power Plant always turns against the wave front. This is a passive way of adjustment (weather vaning) without consuming energy.

The support structure of Poseidon consists of ship-shaped semi-submerged parts. This helps stability as the surface area is distributed, and hence the area moment of inertia increases. The support structure is 75 % submerged (freeboard of 1.5 m and draft of 4.5 m). The characteristic of the Poseidon 37 is listed in Table 6.1.

In theory, the WEC can be of various types, such as point absorber or over-topping devices. Floating Power Plant has chosen a wave-energy device so-called Front Pivot Hinged Wave Absorber for Poseidon hybrid platform. Poseidon has a hydraulic power take-off system. The wave energy part of the concept is a multi ab-sorber system, where the waves force the dynamically ballasted floats up and down. The floats activate hydraulic cylinders that pump water through a water turbine, driving an electric generator (Kallesøe 2011).

Alternative power take-off system can be developed in which the rotational movement of the hinge is converted directly into electricity through a mechanical system or hydraulic pressure to drive a generator and generate electricity. Cut-in and cut-out significant wave heights in Poseidon 37 are 0.2 and 1.5 m, respectively. For a specified offshore site, the design dimensions and thus the cut-in and cut-out values can be varied (lorc.dk 2011).

The wave absorbers (floaters) are hinged at the front. According to Floating Power Plant (FPP), up to 34 % of the incoming wave energy is converted to elec-tricity. This has been confirmed by the research and consultancy organization DHI

Table 6.2 Characteristics of the Gaia-wind turbine. (gaia-wind.com 2014)

Parameter	Value
Rotor diameter	13 m
Tower top mass	900 kg
Tower height	12 m
Rated power	11 kW
Cut-in wind speed	3.5 m/s
Cut-out wind speed	25 m/s
Rated wind speed	9.5 m/s
Rotation speed	56 rpm

(Danish Hydraulic Institute) in the latest wave flume test series (lorc.dk 2011). Each absorber on Poseidon 37 weighs 4.7 t without ballast and 24 t fully ballasted. The ballast system is an active control system that ensures the optimum floater movement: in this way the floaters have high efficiency in small as well as large waves. Trimming of the submerged depth is also actively controlled. Trimming the platform is a part of the optimization process of the wave-energy conversion system (Kallesøe et al. 2009).

Different types and number of wind turbines can be selected by considering the stability and performance. Three 11 kW Gaia wind turbines are mounted on the Poseidon platform, see Table 6.2. The Gaia-turbine has two blades with a teetering hub in which the blades are fixed to each other, but hinge in a bearing such that they can make rigid body rotation out of the rotor plane. Turbine has a free yaw system (a downwind turbine) and runs at a fixed speed, having a gearbox and an asynchronous generator. It has a fixed pitch and uses stall to limit the power production (Kallesøe 2011).

Wave and wind are usually aligned, and this is the optimal scenario for the plant. Misalignment between wave and wind reduces the power production. For crosswinds, the power output from turbines is at its lowest. However, the probability of this environmental condition is low.

The support structure Poseidon is anchored/moored using a Turret Mooring System. Turret mooring is attached to the front part and allows the entire system rotates full 360°. This makes it possible for the structure to turn easily toward waves. Turret mooring system is a standard station-keeping system in the offshore oil and gas technology, which is widely used for FPSO vessels (Floating Production, Storage and Offloading). In practice, the turret is a buoy held in place by three or more mooring lines. The mooring lines are fixed to seabed using anchors, in Poseidon's case, these are plow anchors. A tug boat drags the anchors into the seabed until a specified tension is achieved (lorc.dk 2011).

The mooring lines have enough slack for the turret to move up and down when the water levels rise and fall. However, the horizontal motions remain limited due to stiffness provided by mooring lines. Several mooring lines are usually spread around the buoy and provide enough stiffness in horizontal directions. Thus, platform with turret will be able to follow a rise in the sea level. The Poseidon system is not suited and commercially feasible for short depths, e.g. less than 40 m.

Stability and dynamic performance of Poseidon come from the special design of semi-submerged ship-shaped structure. The surface area is spread and hence the area moment of inertia is increased. The distance between the rear and front parts increases the pitch motion restoring moments. Also, large-surface area at the bottom increase hydrodynamic damping. As it is mentioned, 75 % of the structure is submerged, which increases the metacentric height. The rear section has a triangular shape and acts as a tail rudder. It is seen that the downwind turbine adds more damping in the lateral direction (platform roll). The reason is the free yaw, which changes the characteristic behaviour of the turbine (Kallesøe et al. 2009).

6.4   Synergies of Combined Wave and Wind Concepts

Hybrid offshore energy structures are usually floating or fixed platforms using wind-energy converters combined with an additional wave and/or tidal energy device (However, the focus is wave and wind combination in this chapter). A hybrid concept should be first optimized and standardized based on wind technology practice. As the wind technology is the most mature among the ocean energies. The other technologies, such as wave energy, are non-proven technologies. Therefore, the development of hybrid concepts should be based on wind technology, which is more standardized, mature and proven. Combining with sharing, the substructure is likely to occur after individual devices have proven themselves. Afterward, guarantees, warranties, certifications and insurance should be obtained for the entire combined system. Some of the advantages of combining wave- and wind-energy devices are already mentioned in the previous sections. Here, more explicitly, the advantages and possible disadvantages of hybrid concepts are discussed. Apart from the synergies of co-mounting different devices using same substructure, synergies are classified in three main groups (Casale et al. 2012):

1. Spatial synergies: sharing areas
   The idea is to use the same area for mounting different renewable energy devices and/or for the collocation of different activities, e.g. wind and wave farms.
2. Installation and infrastructure commonalities
   Sharing infrastructures, in particular sharing grid, ports and vessels, seems to be the most promising synergy and the way of cost reduction in short-term scenarios for combined offshore renewable energy systems. Some possible advantageous are listed below:
   - Construction project, in particular if modular structures are considered
   - SCADA and remote control
   - Commonalities in the supply chain: production sites, logistics and storage of components
   - Installation equipments (vessels, jack-ups, cable laying, etc.)

- Port infrastructures
- Grid connection and grid reinforcement
- Storage and storage study
- Operation and maintenance (O&M) synergies

3. Process engineering synergies

Several activities from very different sectors can be combined in principle with offshore renewables, for instance, offshore oil/gas sectors can use the power supply from a nearby wind farm, for more information refer to (Casale 2012).

Co-mounting the offshore wind turbines and WECs provides synergies and several advantages such as (Casale 2012):

1. Energy yield: Co-located technologies would increase the energy yield per unit area of ocean and contribute to a better use of the wind- and wave-energy resources.
2. Grid infrastructure: Sharing the electric grid infrastructure decrease the costs.
3. Logistics: Using common specialized marine equipments needed for the transportation and installation as well as implementing same utilities during the life of the project, such as port space or installation vessels, would also reduce the costs.
4. Operation and maintenance (O&M): Having a common survey, monitoring, repair and maintenance as well as applying common installations and same technicians would become an important cost reduction.
5. Foundation: Sharing the same foundation system decreases the costs of structures compared with separate projects.
6. Shadow effects: Co-locating WECs and Offshore Wind Turbine (OWTs) may have a shadow effect. WECs located on the perimeter of the offshore wind park absorb wave energy and thus result in a milder wave climate inside the park. This effect may open more windows for O&M and can reduce the loads on the OWTs structures; refer to Poseidon project mentioned in the previous sections.
7. Smoothing power output: Waves have less variability and are more predictable than wind, as the wave climate peaks trail the wind peaks. This would help wind parks to avoid the effects of sudden disconnections on the electric grid and to obtain a more accurate output forecast.
8. Environmental benefits: Knowledge of the environmental impacts and metocean data can be transferred among the wave- and wind-energy industry sectors.

The cost savings in combined wave-wind concepts has two main groups: initial and lifetime savings. The initial savings have a direct influence over the capital cost of the project, such as:

- Grid connection: which has a high relative weight, as this is one of the most important costs for an offshore project.
- Licensing: This is the cost reduction linked to licensing of one combined project instead of two projects.
- Substructure: This cost reduction is only for hybrid wave-wind concepts.

Lifetime savings have an effect over the variable cost of a project and are distributed during its life, such as:

- Operation and maintenance (O&M): Sharing the specialized personnel has cost reductions; however, the increased technical complexity means that this reduction will not be extremely effective.
- Weather windows for O&M: Shield effect of the WECs over the combined farm will end on increasing the weather windows for O&M.

There are possible disadvantages when using the same space for both technologies (Pérez and Iglesias 2012; Pérez-Collazo et al. 2013). Some of probable negative effects are listed below:

- Development times: WECs are not mature compared to wind turbines. The early stage of development of WEC technologies involves longer development times which can be an upward factor on the project cost.
- Insurance: The lack of experience and knowledge in co-located hybrid wave- and wind-energy projects can includes higher insurance costs.
- Accident or damage risks: Co-locating floating WECs near OWTs could increase the risk of accident or damage in case of mooring failure of the WEC system.
- Site-selection compromise: Optimizing the site selection for a co-located concept could not be ideal for combined wave and wind energies compared with the stand-alone option.
- Uncertainty of mooring lines: Current mooring lines used in the offshore industry are mostly designed for traditional applications, such as oil and gas. And, their response under the dynamic loading of a WEC, OWT or combination of them has not been refined.
- Failure due to lack of experience: There is a lack of experiences for both arrays of WECs and full-scale prototypes of combined wave-wind systems. Also, there is a lack of real data supporting the reliability of WECs and combined solutions in real conditions.

Pérez-Collazo et al. studied three types of WECs in combination with OWTs (Pérez-Collazo et al. 2013). Three WECs, Wavedragon, Wavebob and Wavestar were considered, for more information about these three types of WECs refer to previous chapters of this book. Special parameters for WECs have been selected which are explained below:

- WEC main active dimension: the main dimension of the WEC, the distance measured in meters for extraction of wave energy which is 260, 15 and 100 m for Wavedragon, Wavebob and Wavestar, correspondingly.
- Capture width: a measure of the performance of the WECs. It measures the ratio between the absorbed energy from the device and the total wave energy available per meter of wave front which is 0.23, 0.42 and 0.40 for Wavedragon, Wavebob and Wavestar, correspondingly.
- Shielding potential coefficient (regular operation): an analytical value to measure the shielding potential of the WECs in standard operational conditions which is 0.6, 0.3 and 0.6 for Wavedragon, Wavebob and Wavestar, correspondingly.

- Shielding potential coefficient (storm operation): an analytical value to measure the shielding potential of the WECs under storm conditions which is 0.5 and 0.2 for Wavedragon and Wavebob, correspondingly.

They have considered three case studies:

a. Co-located wave-wind array where the WECs are distributed between the wind turbines at the periphery of the array facing the incoming waves. Wavebob due to its small size in comparison with the spacing between wind turbines is selected. This approach takes advantage of the shielding capability of the WECs to extract energy from waves and consequently reduces the wave at the inner part of the farm.
b. Hybrid wave-wind-energy converters, the WECs and OWTs are sharing the same foundations. The selected WEC for this hybrid device is Wavestar integrated with a wind turbine.
c. The combination of the solutions mentioned above by introducing small point absorbers at the inner area of the array (using Wavebob), and including the Wavestar hybrid. Also, some periphery wind turbines are removed to install large WEC like Wave Dragon to increase the shielding effect.

The results of analyses showed that (Pérez-Collazo et al. 2013): The case c has the highest saving. Case a has a good cost reduction due to shielding effects. However, it can be concluded that case c has higher risks and case b has small risk. This is due to the point that in case b bottom-fixed hybrid device is applied, which significantly reduces the risks, especially the collision risk.

6.5 Hybrid Wave- and Wind-Energy Concepts

As it is explained above, it is possible to put WECs between offshore wind turbines and use the advantages of sharing CAPEX (Capital Expenditure) and OPEX (Operational Expenditure) between them. This reduces the cost of produced electricity. The combining of wave energy and wind energy can present segregated concepts in which the support-structures used by wave and wind devices are separated. In such case, the main alternative is the segregated bottom-fixed wind turbines combined with WECs. Segregated combinations of floating wind turbines and WECs with either fixed or floating foundations are less feasible and not practical in some conditions.

The simplest approach at the current stage of wave and wind technologies is a co-located independent combination. This type is based on an actual offshore wind farm configuration and deployment of WECs on the same space with sharing common installations and infrastructures like grid connections while having independent foundation systems. Different options can be considered, such as (Pérez and Iglesias 2012):

- Placing the WECs on the perimeter as a wave shield
- Distributing the WECs through the entire wind park
- Using bottom-fixed WECs (to reduce the impact risk)
- Using floating WECs

Table 6.3 Examples of combining wave energy and wind energy, refer to Fig. 6.2

Name	Company, reference	Description
Wave Treader	Green Ocean Energy Ltd, http://www.power-technology.com/projects/greenoceanenergywav/	Bottom-fixed hybrid wave-wind device based on monopile OWT plus point absorber WEC, hydraulic system is applied; the pressurized hydraulic fluid due to movement of WECs is smoothed by hydraulic accumulators before driving a hydraulic motor which drives an electrical generator
2Wave1Wind Platform	Offshore Wind and Wave Energy (OWWE Ltd), http://www.owwe.net/	Wave over-topping devices and point absorbers WECs plus wind turbines
W2 Power Platform	Pelagic Power AS, http://www.pelagicpower.no/	Floating hybrid wave-wind device based on semi-submersible wind turbine plus point absorber buoys as WECs
Poseidon Wave and Wind	Floating Power Plant (FPP), http://www.floatingpower-plant.com/	Floating hybrid wave-wind device, ship-shaped semi-submerged platform with point absorber WECs plus wind turbines
WEGA	Sea for Life, http://www.seaforlife.com	Bottom-fixed hybrid wave-wind device, WEC is a point absorber, an articulated suspended body which is semi-submerged and attached to a mount structure connected to a bottom-fixed wind turbine
WaveCatcher	Offshore Islands Ltd, http://www.offshoreislands-limited.com/	WaveCatcher is a combined point absorber wave- energy device and OWT based on jacket foundations, the original concept has tidal turbines and WECs
Multi Unit Floating Offshore Wind farm (MUFOW)	Barltrop (1993); Henderson et al. (2000)	Floating hybrid wave-wind unit, wind turbines are mounted on a moored floating hull. WECs can be integrated as well
The Langlee E2	Langlee wave power, http://www.langlee.no/	Semi-submersible wave platform which can be integrated in wind parks

The main advantages of segregated bottom-fixed wind turbines and WECs are:

- Simplicity: The concept does not require major changes applied to the current technologies.
- Straightforward integration: deploying the devices on the same area and planning the grid connections and capacity, accordingly.
- Possibility of applying for either bottom-fixed WECs or floating WECs.

There is not much experience in co-located devices, which needs hard efforts by wave or wind developers. The new concepts from floating type signify the risk of accident or collision between the WEC and the OWT. Hence, mooring lines technology should be enhanced. The insurance costs for the global project could be large due to high probability of collision. Some examples of combined wave and wind energies, especially hybrid concepts based on floating and bottom-fixed wind turbines and different types of WECs are listed in Table 6.3. Figure 6.2 shows the concepts.

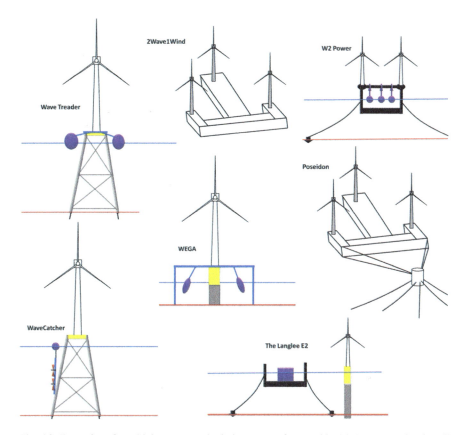

Fig. 6.2 Examples of combining wave and wind energy; refer to Table 6.3. Wave Treader (http://
www.power-technology.com/projects/greenoceanenergywav/), 2Wave1Wind Platform (http://www.
owwe.net/), W2 Power (http://www.pelagicpower.no/), Poseidon (http://www.floatingpowerplant.
com/), WEGA (http://www.seaforlife.com), WaveCatcher (http://www.offshoreislandslimited.
com/), The Langlee E2 (http://www.langlee.no/)

In the following sections, the hybrid energy platforms are discussed, and some
examples of developing concepts around the globe are mentioned. Hundreds of
combinations are imaginable by combining the main offshore wind turbine concepts with main wave-energy convertors. The main points in the selection of a combined system for a specific site are feasibility, serviceability, constructability as well
as the cost of produced energy. For a defined site with specific characteristics and
metocean, just few combinations are practical considering the wave and wind correlation, misalignment, joint probability and intensity. In practice, based on the support structures, it is possible to subdivide the concepts in two main groups: floating
and fixed.

6.6 Bottom-Fixed Hybrid Wind-Wave-Energy Concepts

Hybrid concepts are combined wave- and wind-energy devices in which shared platform/base is implemented. In this section, bottom-fixed hybrid wind-wave energy devices are discussed. This type is based on bottom-fixed offshore wind technology to make a base for integrating wave energy and creating new hybrid devices. Limited modifications of the foundation/support structure of wind turbine are needed for adapting an already existent wave device. In order to limit the extra cost and deviation from original design of the offshore wind turbine, minimum changes of support structure should be applied.

Several hybrid concepts based on original bottom-fixed wind turbines (e.g. gravity-based, monopile, jacket, tripod and tripile) and original WECs (oscillating water column, point absorber and over-topping) are imaginable. However, deep studies and researches are needed to investigate the feasibility of combining a bottom-fixed wind turbine and a WEC. Feasibility of such a hybrid concept is related to a specific offshore site, such as considering water depth, metocean, wave and wind resources, joint probability and misalignment of wave and wind.

The simplest option is a hybrid concept based on monopile wind turbine and point absorber WEC. Figure 6.3 illustrates a bottom-fixed hybrid wave-wind concept. In such a concept, floating buoys, e.g. 2–3, are located around the monopile wind turbine. The buoys are moving up and down as wave passes; the energy of the waves is transferred to the power take-off system (e.g. a hydraulic type in this case). The buoys are supported by load-carrying beams which connect the buoys to the support structure. The number and dimensions of the buoys can be optimized considering the wave-energy resource at the site, environmental loads and structural integrity of the entire hybrid system. One of the main issues is the extra loading from the waves on the support structures, which should be precisely analysed to investigate extra cost due to needed strengthening. The extra power produced with the hybrid concept and gained synergies should be compared to the extra cost to see the feasibility of a specific concept in a long-term period, e.g. in a 25 years design life.

The advantages of the bottom-fixed hybrid wind-wave energy are (Pérez and Iglesias 2012):

- Lack of floating bodies and hence reducing the risk of accident and consequently the insurance cost
- Cost reduction due to the use of a shared foundation system
- Increase in the energy yield per device

The disadvantages of the bottom-fixed hybrid wind-wave energy are:

- Extra costs for production and installation of the foundation system
- Extra loads on the structure

The bottom-fixed wind turbines can easily serve as a foundation for WECs. Example of combining a monopile wind turbine with a point absorbing WEC is shown

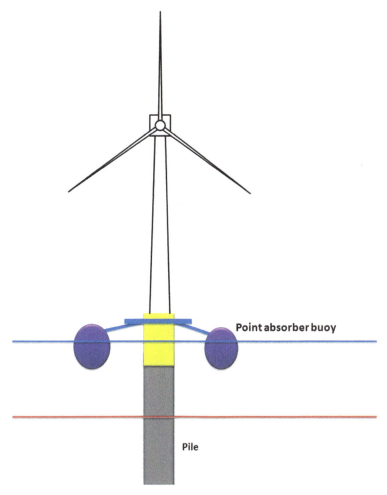

Fig. 6.3 Schematic layout of a bottom-fixed hybrid wave-wind-energy concept

in Fig. 6.4. A torus point absorber is located around the monopile. The torus buoy is connected to hydraulic actuators fixed on the foundation structure. Wave oscillations move the torus up/down, transmitting the movement to the hydraulic actuators. The high-pressure oil from the hydraulic actuators drives a hydraulic generator to make electrical power.

In the design of hybrid wave-wind-energy concepts, lightness of the entire system is an advantage. Torus configuration is relatively simple and light structure. Hydraulic system can be an option for generating electricity. However, in general, hydraulic systems used in WECs have a high operation and maintenance (O&M) costs. WECs are still at the development stage, and more research is necessary to achieve higher performances. The generator and hydraulic actuators need a support

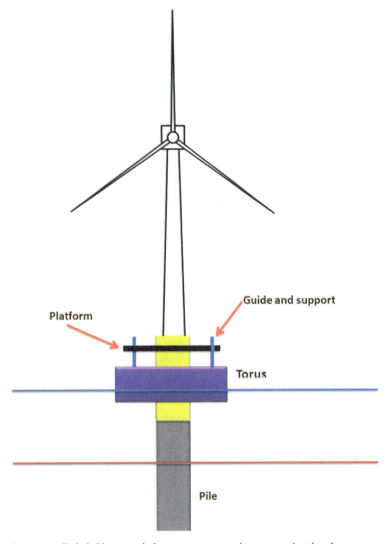

Fig. 6.4 A monopile hybrid wave-wind-energy concept using torus point absorber

structure fixed to wind turbine, see Fig. 6.4. The platform can facilitate access to the wind turbine and wave-energy device for monitoring, repair and maintenance operations.

The produced power of WEC compared to OWT is much smaller, e.g. just 10%. However, there are other synergies, such as a shared cost of substructure, gird and smoother power, which motivate us for integrating wave and wind on a same unit. Another parameter that should be considered in the design is the relative ratio of the exposed area to wave for WEC compared to OWT (e.g. the diameter of the torus compared to the diameter of the monopile). To harness wave energy, higher width of

WEC is generally needed. This means that the structure is subjected to more wave loads. On the other hand, an increased wave load on the structure is a negative effect for OWT. This is a paradox in the design of hybrid concepts. The final decision is made based on the optimized size and shape of the WEC considering the entire structure integrity, total wave-wind produced electricity and costs. Hence, the entire system should be investigated as a unit subjected to wave and wind loads in coupled time domain analyses.

Wave Treader Scottish company Green Ocean Energy is in the process of developing the Wave Treader machine. The Wave Treader prototype will be tested in the UK. Wave Treader is primarily aimed for the UK round three offshore wind farms and Scottish territorial waters. Around 7500–8300 offshore wind turbines are expected to be installed in the UK between 2015 and 2023 (power-technology.com).

The Wave Treader is a point absorber WEC that shares the offshore wind-farm infrastructure and helps increasing yield from the same ocean surface. It is attached to the structural part of offshore wind turbine at the mean water surface, e.g. at the transition piece of a monopile. It can ease the access to the turbine for survey and maintenance and repair. Also, it increases the power production of the farm while smoothing the produced electricity. It comprises of steel braces and two floating bodies molded of glass-reinforced plastic (GRP).

Standard hydraulic and electrical equipments are applied and are basically "off-the-shelf" items. Wave Treader is currently designed to have 500–700 kW of peak rating. The modular design of the machine will make it easier to change components; the operational life is 25 years with refits every 5 years.

Wave energy powers arms of the Wave Treader to move up and down, which stroke their hydraulic cylinder. The hydraulic cylinder drives the electric generator. The produced electricity is exported through the wind turbine power cables. The interface structure moves vertically in the tidal range and also rotates to ensure optimal alignment to the direction of the waves. Wave Treader can also be fitted to the already existing wind farms (power-technology.com).

Wavestar The Wavestar 5 MW system can produce each kWh of electricity at 8 cents. This is planned to be reduced in future as low as 4.7 cents for each kWh and reach the offshore wind-power price, refer to (Steenstrup 2006). Wavestar can be combined with bottom-fixed wind turbines. As the WEC is installed on the seabed using the same monopile technology as applied for wind turbines, the Wavestar WEC can be placed inside the wind farm. Theoretically, up to five WECs can be placed between each wind turbine on every other row. The OWT and WECs are sharing the same grid and common foundation (Marquis et al. 2012). DONG Energy A/S and Wavestar A/S have started a research and development collaboration to explore the prospects of combining the wind and wave power.

In a longer period, a hybrid concept integrating Wavestar with bottom-fixed wind turbine may appear, see Fig. 6.5. A 5 MW wind turbine can be implemented on the wave device structure comprising three WECs placed on a star combination with a capacity of 2.4 MW. The complete hybrid platform will have a total capacity of 7.4 MW in this way.

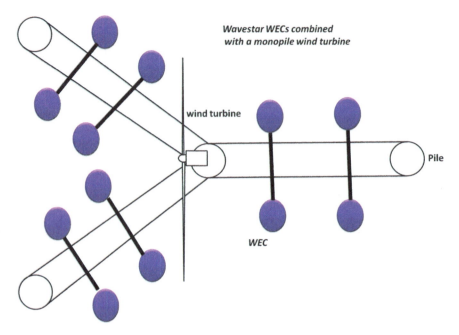

Fig. 6.5 Wavestar wave-energy converters (*WECs*) combined with a monopile wind turbine (top view). (wavestarenergy.com 2014)

6.7   Floating Hybrid Wind-Wave-Energy Concepts

As mentioned before, both WEC technology and floating wind turbine expertise are in the developing stage. WECs need more enhancement and development, and floating wind turbines are not mature compared to bottom-fixed wind turbines. In the long term, combining WECs and Floating Wind Turbine (FWTs) together in hybrid configuration should be considered. Currently, there are different research groups and universities studying the feasibility of floating hybrid wave-wind concepts.

The main advantage of hybrid floating solutions is the possibility to reach deep-water wave and wind resources. The majority of wave resources of the European region is located in deep-water area, e.g. at the Atlantic Ocean. Floating hybrid concepts are at the early development stage, and they may appear in future. In this chapter, the main floating hybrid wave-wind devices and the current status of the existing research concepts around the world are discussed.

The idea is developing hybrid floating concepts that integrate WECs and FWTs in a hybrid platform. There are two possibilities to make a floating hybrid wind-wave concept from WECs and OWTs. Either mounting wind turbines on big WECs, e.g. big overtopping WECs could be used as a floating foundation system for one or several smaller wind turbines. Or, integrating small WECs on a floating offshore wind turbine substructure, e.g. as part of the system to reduce the wave-wind-induced motions of the floating turbine by extracting the wave energy. The power take off appears as damper in this way.

Fig. 6.6 WindWaveFloat, a floating hybrid wave-wind concept based on semi-submersible wind turbine integrated with point absorber WEC, original picture. (Source: commons. wikimedia.org (Untrakdrover 2012), the picture is modified by the author to include the WEC in the middle of the platform. The file is licensed under the Creative Commons Attribution-Share Alike 3.0 Unported license)

An example of the later type is a combination of semi-submersible wind turbine with point absorber WECs. Principle Power is developing an innovative technology with the potential to generate electricity from the winds and waves. The hybrid marine platform, so-called WindWaveFloat, will combine the floating offshore wind turbine platform with wave-energy convertors, so the system can simultaneously generate electricity from both winds and waves. The WindFloat concept is discussed in the previous chapters. For more information regarding the hybrid "WindWaveFloat" device, refer to (Higgins 2011).

WindWaveFloat Antoine Peiffer and Dominique Roddier documented the design of a hybrid device based on combining a WEC on the WindFloat structure. Their paper summarized the numerical modeling and experimental testing that were performed to integrate an Oscillating Wave Surge Converter (OWSC) on the Wind-Float structure (Peiffer and Roddier 2012). More studies have been carried out to investigate the possible solutions adding different WECs to WindFloat. Oscillating water columns, buoy point absorbers, spherical-point absorber as well as flaps have been investigated; refer to (Weinstein 2011). In general, this hybrid floating wave-wind concept integrating the WECs and WindFloat semi-submersible wind turbine is called "WindWaveFloat". Figure 6.6 shows WindWaveFloat based on spherical single-point energy absorber. The results from the studies carried out by Principle power for investigating the feasibility of WindWaveFloat is summarized here.

1. Oscillating Water Column

The WindFloat and WindWaveFloat are based on semi-submersible floater having three columns. The first column carried the wind turbine. The oscillating water columns (OWCs) are integrated in column 2 and 3 by creating a chamber around the columns. OWCs have a proven technology, and they are robust. The compressed air runs through a wells turbine. The challenges for such a concept are that the significant wave loading and efficiency losses can be significant.

2. Spherical Wave-Energy Device (SWEDE)

A single-point energy absorber with spherical shape is connected to all three columns. The sphere floater movement (e.g. in heave/sways/surge) due to wave loads generates electricity. The challenges of such a concept are deign failure mode when large floater is applied as well as accommodating the floater inside the WindFloat and hydrodynamic interactions.

3. Oscillating Vertical Plates (flaps)

Three flat plates (flaps) oscillating around the main beams between the columns can be implemented. The torque due to wave loads on flaps drive the PTO. In harsh conditions, it is possible to set the flaps out of waves for survival conditions by simply rotating them. The challenge is the structural integrity of the braces connected to flaps and taking the extra wave loads at joints considering the fatigue limit states.

4. Point Absorbers

Independent point absorbers, e.g. heaving buoy can be implemented.

   The experiments and numerical analyses showed that (Weinstein 2011) most wave-energy conversion (PTOs) hardly affected the motions of the WindFloat platform. And, consequently, the power production and integrity of the system are not significantly affected. However, more extensive analysis required for the design of hybrid marine platforms. The study had some assumptions, which can be improved, such as the efficiency of the OWC system. In future, the geometry of the wave-energy device can be optimized to improve its performance.

W2Power W2Power is a floating hybrid wave-wind concept based on integrating wind turbines supported on a semi-submersible together with WECs (point absorber type), see Fig. 6.7. The concept is being developed for moderate and deep water to harness the wind and wave energies. The combination of wind and wave power in W2Power and similar hybrid concepts helps to meet the energy demands. As few countries have extensive shallow seas available for wind development. W2Power floating hybrid wind and wave-energy conversion plant consists of two wind turbines mounted on a semi-submersible triangle support structure. The third corner houses the wave energy power take-off using a Pelton turbine (Pelton turbine is widely applied in hydropower applications). The wave-conversion technology is based on wave-driven seawater pumps. The Pelton turbine is driven by three lines of wave-actuated pumps mounted on the platform's sides.

   Two 3.6 MW standard offshore wind turbines, e.g. the Siemens 3.6–107 (107 m rotor diameter and hub height of 80–85 m) can be applied. This hybrid marine

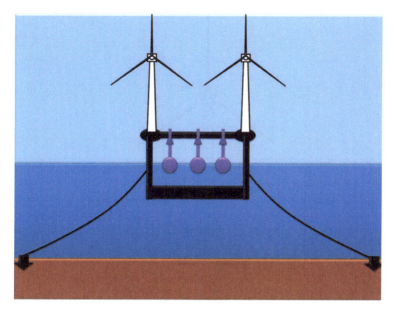

Fig. 6.7 Schematic layout of W2Power, a floating hybrid wave-wind concept based on semi-submersible wind turbine integrated with point absorber WECs (pelagicpower.no/, 2010). (pelagicpower.no/ 2010)

platform may be rated at more than 10 MW for offshore areas with a strong wave climate (pelagicpower.no/ 2010). By using counter-rotating wind turbines, the side-way forces can be effectively reduced. The thrust and the gyro-force from the WT's may contribute to stabilize the platform. The turret mooring design allows the platform to yaw and eliminate the need for individual turbine yawing.

Spar Torus Combination (STC), a Patented Hybrid Concept Professor Torgeir Moan, Dr. Madjid Karimirad, Dr. Zhen Gao and Dr. Made Jaya Muliawan patented (CeSOS 2011) an efficient floating hybrid concept based on integrating spar-type floating wind turbine with torus point absorber WEC so-called STC, refer to (Muliawan et al. 2013b). Figure 6.8 illustrates the schematic layout of the STC concept, a floating hybrid wave-wind device for deep-water areas. The concept is patented (Patent publication number: WO 2013137744 A1) and several international publications are available during the past years. The idea was born in CeSOS/NTNU several years ago, and it is continuously being developed by the research team. STC has been recently selected by Marina EU project (http://www.marina-platform.info/) as one of the focusing concepts among hundreds of proposed hybrid concepts.

STC is a combined concept involving a combination of spar-type FWTs and an axi-symmetric two-body WECs. Compared with segregated deployments of FWTs and WECs, STC would imply reduced capital costs of the total project because it will reduce the number of power cables, mooring line and the structural mass of the WECs, refer to (Muliawan et al. 2013a).

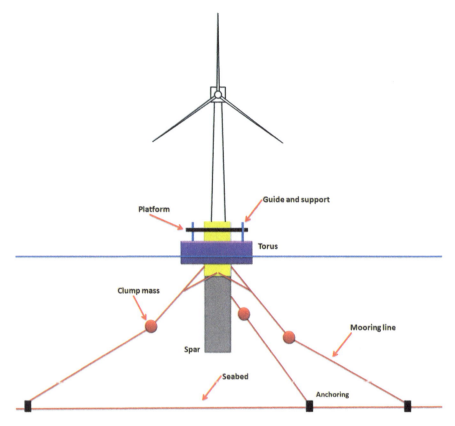

Fig. 6.8 STC, spar torus combination, a floating hybrid wave-wind concept based on spar-type wind turbine integrated with point absorber WEC; patent publication number: WO 2013137744 A1; refer to CeSOS (2011)

For STC, a torus (donut-shape heaving buoy) with a spar-type FWT, an insignificant effect due to the addition of a torus is observed for the surge/pitch motions of the spar. However, in the heave mode, the torus buoy follows the water level, and it carries the spar. The addition of a torus on the spar-type FWT increases the mean displacement of the spar slightly but decreases the standard deviation of the spar's motions, especially for surge and pitch responses. This is because this specific torus damps the spar's motions. As a consequence of more stable pitch motion, the WT in the STC concept will get better exposure to the incoming wind, therefore experience higher aerodynamic load to produce electrical power. In particular, compared to spar-type FWT at wind speed lower than the rated wind speed, the power production is higher. The estimated total power production by the combined concept is 10–15 % higher than the one produced by the specified spar-type FWT alone. All of the results indicate that the presented combined concept not only reduces the total capital cost but also increases the total power production compared to those for segregated FWT and WEC concepts (Muliawan et al. 2013a).

References

Barltrop, N. (1993). Multiple unit floating offshore wind farm (MUFOW). *Wind Engineering, 17*(4), 183–188.

Casale, C., Serri, L., RSE, Stolk, N., Yildiz I, Ecofys, Cantù, M., & ENEL Ingegneria e Innovazione. (2012). *Synergies, innovative designs and concepts for multipurpose use of conversion platforms.* Results of ORECCA Project—WP4. http://www.orecca.eu/project/wp4. Accessed Nov 2014.

CeSOS. (2011). *Patent No. WO 2013137744 A1.* Norway.

Chozas, J. F. (2012). *CO-PRODUCTION of WAVE and WIND POWER and its INTEGRATION into ELECTRICITY MARKETS: Case study: Wavestar and 525 kW turbine.* The annual symposium of INORE, the international network on offshore renewable energy. Denmark.

Chozas, J. F., Sørensen, H. C., & Helstrup, N. E. J. (2012). *Economic benefit of combining wave and wind power productions in day-ahead electricity markets.* 4th International Conference on Ocean Energy. Dublin.

Churchill, S. (2011). *Floating power.* http://portlandwiki.org/File:Floating.power.jpg.

floatingpowerplant.com. (2012). *Poseidon floating power.* http://www.floatingpowerplant.com/, http://mhk.pnnl.gov/wiki/index.php/File:Poseidon1.png.

Fusco, F., Nolan, G., & Ringwood, J. V. (2010). Variability reduction through optimal combination of wind/wave resources—An Irish case study. *Energy, 35*(1), 314–325.

Henderson, A., Patel, M. H., Halliday, J., & Watson, G. (2000). *Floating offshore wind farms—an option?* Offshore Wind Energy in Mediterranean and Other European Seas (OWEMES). Italy.

Higgins, M. (2011). *Innovative deepwater platform aims to harness offshore wind and wave power.* http://energy.gov/articles/innovative-deepwater-platform-aims-harness-offshore-wind-and-wave-power.

Kallesøe, B. S. (2011). *Aero-hydro-elastic simulation platform for wave energy systems and floating wind turbines.* Denmark: DTU, Risø-R-1767(EN).

Kallesøe, B. S., Dixen, F. H., Hansen, H. F., & Køhler, A. (2009). *Prototype test and modeling of a combined wave and wind energy conversion system.* Proceedings of the 8th European wave and tidal energy conference. Uppsala, Sweden.

Lakkoju, V. N. (1996). Combined power generation with wind and ocean waves. World renewable energy congress. Denver, USA.

lorc.dk. (2011). *When floating structures combine wind and wave.* http://www.lorc.dk/offshore-wind/foundations/floating-structure-combining-wind-and-wave.

Marquis, L., Kramer, M. M., Kringelum, J., Chozas, J. F., & Helstrup, N. E. (2012). Introduction of wavestar wave energy converter at the danish offshore wind Power Plant Horns Rev 2. ICOE. Dublin. http://www.icoe2012dublin.com.

Muliawan, M. J., Karimirad, M., & Moan, T. (2013a). Dynamic response and power performance of a combined spar-type floating wind turbine and coaxial floating wave energy converter. *Renewable Energy, 50,* 47–57.

Muliawan, M. J., Karimirad, M., Gao, Z., & Moan, T. (2013b). Extreme responses of a combined spar-type floating wind turbine and floating wave energy converter (STC) system with survival modes. *Ocean Engineering, 65,* 71–82.

Peiffer, A. & Roddier, D. (2012). *Design of an oscillating wave surge converter on the windfloat structure.* 4th International Conference on Ocean Energy. Dublin: ICOE2012.

pelagicpower.no/. (2010). W2Power. http://pelagicpower.no/.

Pérez, C., & Iglesias, G. (2012). *Integration of wave energy converters and offshore windmills.* 4th International Conference on Ocean Energy. Dublin: ICOE.

Pérez-Collazo, C., Jakobsen, M. M., Buckland, H., & Fernández-Chozas, J. (2013). Synergies for a wave-wind energy concept. EWEA, Offshore-2013.

Steenstrup, P. R. (2006). Wave Star—A wave power generating system aiming at commercialisation. WAVE STAR—Wave power system. ftp://ftp.cordis.europa.eu/pub/sustdev/docs/energy/resen_steenstrup.pdf.

Stoutenburg, E. D., & Jacobson, M. Z. (2010). *Optimizing offshore transmission links for marine renewable energy farms*. USA: Stanford Univeristy.

Untrakdrover. (2012). Agucadoura windfloat prototype. http://commons.wikimedia.org/wiki/File:Agucadoura_WindFloat_Prototype.jpg.

wavestarenergy.com. (2014). Wave star energy.

Weinstein, A. (2011). WindWaveFloat. http://www1.eere.energy.gov/water/pdfs/11_wwf_principle_power_weinstein.pdf.

Chapter 7
Design Aspects

7.1 Introduction

In the previous chapters, the wind turbines, wave-energy converters, combined wave and wind energy as well as hybrid marine platforms are explained. This chapter focuses on the design aspects of marine structures. Offshore renewable energy devices are facing the marine environments and should survive the ocean wave and wind loads during their life. Hence, similar to the other marine structures, such as oil/gas platforms, ships and coastal structures, the offshore energy structures need to comply with especial design requirements governed by standards.

Offshore oil/gas platforms are usually designed and built one-of-a-kind for a specific offshore site while offshore energy structures are usually set in an array consisting of 10–100 units. Currently, the state-of-the-art for offshore energy structures, in principle, is engineered for site-specific design. However, to advance the offshore energy applications, it is needed to have custom-made type-approved designs. This considers mass production of WECs, FWTs, OWTs and hybrid marine platforms in near future. Moreover, the structural design should not be for a specific site, but rather for a class of environmental conditions. The idea can be to have approved designs and concepts for a range of water depths and metocean. The designs should cover sets of support structures, foundations, wave power take off and wind-energy devices.

7.2 What is Design?

To start, it is a good point to remember a well-known quote by Professor Albert Einstein (1879–1955): "The best design is the simplest one that works."

If we think about this quote, it is clear that the design means a product that should properly work with minimum overall-cost while maintaining an acceptable level of

© Springer International Publishing Switzerland 2014
M. Karimirad, *Offshore Energy Structures,* DOI 10.1007/978-3-319-12175-8_7

safety with adequate reliability when it simply continues to work. More specifically, a design is a product of a well-established process in which:

- Acceptable level of safety by setting requirements for structures and structural components is achieved.
- Needed references are usually made to standards, recommended practices, guidelines and rules.
- All phases of structure life, including engineering, construction, installation and in-service (in-place) cases are considered.

In the following sections, design of offshore energy structures is discussed in detail. However, let us list the main items relevant for such a design to have an idea about the subjects, which should be covered in this chapter. Simply, a design of an offshore energy structure should at least cover the following items among the others:

- Design principles mentioned in standards, regulations and rules
- Material properties and procurement considering inspection in the yard
- Offshore site specification and metocean data
- Load cases, design load calculations, and load effect analyses
- Load combinations and safety factors
- Limit states and design criteria
- Structural strength and component design
- Foundation and mooring systems (station keeping)
- Anchoring and fairleads (both temporary and permanent)
- Corrosion protection, marine growth and fouling
- Marine operations, including transport and installation
- In-place considerations, survey and inspection
- Subsea infrastructure, electric grid and power cable design

Some of the items listed above are directly discussed in this chapter while the others are covered in the following chapters.

7.3   General Design Aspects

The design of offshore structures should reflect several issues including a long-term perspective of engineering, procurement, construction, installation, maintenance, monitoring and operation. Structures and structural components (members and elements, both load carrying and those that are functioning without taking a significant loading) are designed to

- Withstand the loading during temporary, operating, intact and damaged conditions
- Have acceptable safety during the design life (e.g. for 25 years)
- Provide acceptable safety for personnel and environment
- Ensure sufficient robustness against deterioration during the design life

Also, the fabrication/construction of the structure and its components should comply with recognized techniques and practices. The design should ease the construction, transportation and installation as far as it is possible. The inspection, maintenance and repair during the design life should be considered as well. Structure and structural parts must have enough ductile resistance, and the design must minimize stress concentrations and reduce complex stress-flow patterns. This chapter briefly highlights important points in designing of offshore renewable energy structures. Reliability, economy and environmental aspects are the key issues.

Offshore site and metocean conditions have a great influence on the design. The following items are needed prior to start an offshore project.

• Subsea conditions (current, wave kinematics, wave–current interactions)
• Water depth and tidal range
• Sea surface conditions (wave spectra, directionality/spreading, wind)

7.4   Reliability and Limit States

To design a safe and robust structure, the reliability of the system subjected to loads should be considered. Analyses requirements for checking the structural integrity of the structure for different limit states should be reflected in the design. A limit state is a condition beyond which a structure or structural component will no longer satisfy the design requirements. Limit states and safety classes are defined to document the strength and performance of the system in operation and harsh conditions. Fatigue limit state (FLS), ultimate limit state (ULS), accidental limit state (ALS), serviceability limit state (SLS) are the main limit states defined for marine structures (DNV 2013b Design of offshore wind turbine structures).

• ULS considers the maximum load-carrying resistance

   – Loss of structural resistance under yielding and buckling
   – Brittle fracture of elements
   – Loss of static equilibrium of the structure, e.g. overturning or capsizing
   – Failure of critical components
   – Excessive deformation of the components

• FLS considers failures under cyclic loading, e.g. cumulative damage due to repeated loads
• ALS considers either (1) maximum load-carrying capacity for accidental events or (2) post-accidental integrity for damaged structures

   – Structural damage caused by accidental loads (type 1)
   – Ultimate resistance of damaged structures (type 2)
   – Loss of structural integrity after local damage (type 2)

- SLS considers criteria/tolerances under normal use

  - Deflections and deformations that modify the acting forces or distribution of loads
  - Excessive vibrations producing discomfort or affecting non-structural components
  - Motions that exceed the limitation of equipment
  - Temperature-induced deformations

Few examples of limit states for offshore energy structures are given below:

- ULS: extreme structural load effects in harsh conditions for a floating wind turbine, maximum responses of a bottom-fixed turbine in environmental conditions associated with rated-wind speed, static stability of a jacket wind turbine sitting on the seafloor before driving piles, failure of braces of a semisubmersible wind turbine under slamming loads in severe sea states
- FLS: cumulative fatigue damage of a structural member, e.g. a joint of a jacket-supported wave-wind energy device
- ALS: collapse of a monopile wind turbine due to ship impact, progressive collapse of a jacket wind turbine due to the failure of one of its critical braces due to fatigue, the capacity of a semisubmersible wind turbine under wave loads after damage due to accidents
- SLS: nacelle-rotor assembly accelerations, differential settlements of foundation soil causing intolerable tilt of the wind turbine, vibrations affecting the controller or electrical components

In general, the limit states are individually considered. However, for a structure in real life over the years, the limit states are linked. For example, the fatigue damage of the members reduces their capacity and consequently, the overall capacity of the structure is affected. Hence, the ULS and ALS may be affected by FLS, e.g. the ultimate strength of the aged structure due to fatigue is less.

Offshore renewable energy structures are usually unmanned and hence, different safety classes can be applied for such structures. Relevant consequence classes for unmanned equipment should be defined as well. Still, there are several discussions and debates among specialist around the world to assign proper safety classes and reliability targets for offshore renewable structures. The problem gets difficult as there are instances of time, such as surveying/repair that a needy human presence at the site. This means even the marine renewable structures are basically unmanned, however, in some cases, human presence should be considered. This requires proper investigation of safety. Human may be on board of such structures while they are designed assuming to be unmanned.

To improve the reliability of offshore energy structures, insight from documented failure examples helps a lot. Knowledge about the mean time before the failure of components in real life enhances the design. Unfortunately, there is a lack of field experience and condition monitoring for offshore renewable energy structures. Also, the performance uncertainties about using conventional components in offshore renewable applications are not properly documented. Documentation of failures and performance of used conventional components in marine renewable applications can advance innovative design.

7.5 Economical Aspects of Design

Capital expenditure (CAPEX) of system and components is a key factor in the development of offshore renewables. The cost of electricity is tightly related to capital cost used for the construction of the power units. New ways of reducing the CAPEX through optimizations of the system considering effective and robust components are needed to make the offshore renewable industry enough mature. Same as CAPEX, the operational expenditure (OPEX) has an important role in the actual price of generated electrical power of an offshore renewable unit. Sensitivity studies considering the operation/maintenance costs and making a relation to reliability are important to find new solutions to decrease the costs. The optimal solution should consider both CAPEX and OPEX in one picture. Note that, generally, when the CAPEX increases, the OPEX reduces as the reliability of the components and design is greater. Other aspects (among other issues), such as the following should be considered when optimizing the design with respect to cost.

- Transportation
- Installation
- Maintenance considering the accessibility and availability
- Decommissioning and partial decommissioning
- Distance to port
- Weather window and environmental conditions at the site
- Access space
- Time and vessel required for installation and repair
- Redundancy and its effects (reducing or increasing) on the cost for an acceptable integrity
- Design class
- Impact of damage to a specific component

7.6 Environmental Aspects of Design

Renewable energies are well-known as green energy, helping to have a clean environment by reducing the air pollution. Hence, it is not favour to lose this advantage by affecting the environment. Studies should concern the possible side effects of the offshore renewables on the ocean and creators living inside or close to it. The following aspects should be considered at least:

- Interaction with marine species
- Component and vessel impact (e.g. emissions)

- Noise
- Scour
- Mooring/foundation footprint radius and imprint
- Electromagnetic effects
- Chemical contamination
- Nursery sites for marine species
- Impact on other water users

7.7   Component Design

As it is mentioned in the previous chapters, offshore energy structures comprise of several components. Figure 7.1 shows a tripile wind turbine components.

In this chapter, the main parts, such as mooring lines and foundations are discussed. In a wind-wave energy farm, there are more components, such as electrical cable, gird and infrastructure, etc. But the design aspects of the following parts are mentioned herein.

- Foundation: connects/stabilizes the support structure to the sea bottom
- Anchoring: connects the mooring lines to seabed in the case of floating structures
- Mooring lines: keep the support structure in a position with a small deviation due to system and environmental loads such as wave/current/wind loads
- Support structure: is the base for the main structure, e.g. supports the wind turbine as a topside
- Wind turbine: consists of a tower and rotor/nacelle assembly, in general
- WEC: wave energy converter, is a device converting the wave energy to mechanical/hydraulic power.
- PTO: power take off system can be incorporated out of water (e.g. mounted on a small platform connected to the support structure) or inside of water (e.g. at the seabed)

A) Design Aspects of Foundation Foundations used in offshore energy structures are based on offshore oil/gas business and marine industry experiences. This includes different types such as:

- Gravity based (GBS)
- Suction
- Embedment
- Sand/rock screws
- Piles
- Novel designs, e.g. caissons either in the form of monopod or tripod/tetrapod arrangement

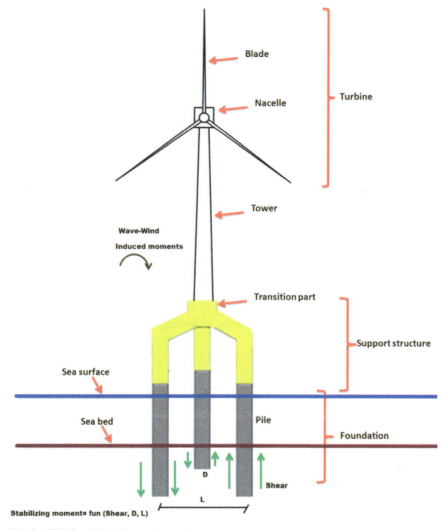

Fig. 7.1 Tripile wind turbine components

The foundation's design criteria should consider the following aspects among the other relevant issues.

- Subsea infrastructure integration
- Grouts design considering possible failures due to wash-out during installation and scour
- Installation requirements considering weather window, time and required equipment
- Corrosion
- Handling and local availability

- Design-load cases
- Dynamics considering

  - Geotechnical tools
  - Fracture/fatigue
  - Scour
  - Grouting processes

- Holding capacity
- Directionality
- Shared or single attachment points
- Environmental and metocean
- Seabed conditions
- Mapping resolution and availability
- Geotechnical considerations (seabed type and variation, ability to withstand dynamic loading)

B) Design Aspects of Anchoring and Mooring System A conventional mooring system based on experiences in offshore technology is currently used. However, new design requirements considering characteristics and properties of offshore energy structures are being developed. Different types of mooring have been widely applied in marine technology including the following:

- Catenary
- Taut
- Single point
- Turret
- Semi-taut

In designing of mooring system, the following issues among the others are important:

- Handling
- Mass and buoyancy: clump masses and buoyancy elements
- System stiffness and damping
- Component temporal changes (damage, fatigue, creep)
- Number of lines
- Pre-tension
- Material:

  - Chain
  - Steel components
  - Synthetics
  - Polyester
  - Nylon
  - Elastomer
  - Spring/damper
  - Articulated leg
  - Composite

- Biofouling, marine growth
- Degradation/corrosion
- Role

  - Station-keeping (avoiding resonance)
  - Part of PTO (designing for resonance)

- Spatial requirements

  - Array topology/ Separation distance
  - Weather-warning
  - Footprint

- Design load cases and analyses

  - First- and second-order wave loading
  - Vessel impact
  - Device/line motion and current
  - High frequency effects (e.g. VIV)
  - Mean drift loads
  - Quasi-static analysis
  - Coupled dynamic (RAO motions/forces, Hydrodynamic parameters, Seabed conditions, Foundation representation, Power take-off representation)
  - Component tools
  - Fatigue and breaking tension (ULS)

- Umbilical constraints

C) Design Aspects of Support Structure Support-structures of renewable offshore energy designs are usually steel marine structures. Offshore oil and gas industry has years of experience in designing, engineering, construction, transportation and installation of platforms. The theories, codes and standards applied in the designing of such structures are verified and validated against experimental data, sea trail and full-scale measurements during the past century. Lots of platforms were built, used and decommissioned afterwards in the oil/gas industry. Hence, this industry has a solid backbone, which supports the new designs.

When it comes to renewable offshore energy structures, there is a lack of experience, and just a small number of concepts have been built. In this situation, it is necessary to start with current practice applied in oil/gas business until custom-made standards, codes and theories appear. This may end up with overdesign and conservative solutions in the first place, which can be optimized and slightly modified based on technical experiences. In long term, reliability methods covering the system and components should be applied to find new safety factors necessary for offshore renewables. This necessitates research and experiments of full-scale units, farms and arrays of offshore renewables. A database covering failures of components, downtimes, faults and reasons of failures should be gathered. Currently, scientists and engineers are investigating the above-mentioned aspects and several joint projects are defined to fill the technical/knowledge gaps. Still, the knowledge

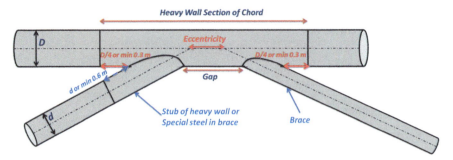

Fig. 7.2 Simple joint details. (For more information on joints refer to NORSOK 2004a Design of steel structures)

core will be offshore oil/gas practice for some years, until reliable and tested approaches, codes, recommendations and standards globally accepted appear. Here, an example is given to show the requirements for knowledge developments for offshore wind technology.

As it is already mentioned, for both bottom-fixed and floating wind turbines, braces may be used as load-carrying members, e.g. in jacket and tripod wind turbines or semisubmersible wind turbine. Where two or more members connect (e.g. the locations where braces meet legs) is called joint. A joint is sensitive to fatigue and lots of studies, both numerical and experimental, have been performed to derive the stress-concentration factor (SCF). There are some design requirements with respect to joint geometry. The gap for simple K-joints should be larger than 50 mm and less than D, see Fig. 7.2. Minimum distances for chord cans and brace stubs should not include thickness tapers (refer, e.g. to NORSOK standard; NORSOK 2004a Design of steel structures). The member and joint strength should be checked with respect to structural integrity. Strength formula of simple tubular joints exists. Joint classification is the process whereby the axial force in a given brace is subdivided into K, X and Y components of actions; corresponding to the three joint types, resistance equations exist, see Fig. 7.3.

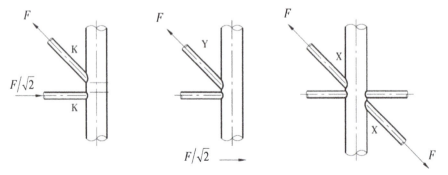

Fig. 7.3 Simple joint classes. (For more information regarding joint classes refer to NORSOK 2004a Design of steel structures)

Members of the jacket/tripod or semisubmersible offshore wind turbines have different dimensions from oil/gas platforms, and hence, their joint classification as well as stress-concentration factors could be different. When there is not enough information regarding the stress-concentration factors, finite element analysis should be applied to investigate the stress pattern for joints. In future, more experiments and analysis are needed to drive proper joint classes and strength analysis for dimensions of braces and joints applied for offshore renewable energies. Structural members of such energy structures are subjected to both dynamic wave and wind loads, which can result in different strength analysis and requirements.

In designing of support structures, the following issues should be considered in general (DNV 2013b "Design of Offshore Wind turbine Structures"):

- Structural components and details should be shaped such that the structure behaves in the presumed ductile manner
- Connections should be designed with smooth transitions and proper alignment of elements
- Stress concentrations should be avoided as far as possible
- Sudden changes in section properties should be avoided

Considering a proper level of safety, the structural dimensions may be reduced if the design is based on an assumption of inspections, maintenance and repair throughout its design life, and the structural load effects are governed by FLS. It may be difficult to apply this to designs governed by ULS. For example, a jacket wind turbine (bottom-fixed) is likely to be governed by FLS and a spar wind turbine (floating) is likely to be governed by ULS depending on offshore site conditions and metocean.

7.8 Design Principles

Different design methodologies and principles exist for offshore structures. DNV categorized them in the following classes (DNV 2013b "Design of Offshore Wind Turbine Structures"):

- Partial safety factor method with linear combination of loads or load effects
- Partial safety factor method with direct simulation of combined load effects of simultaneous load processes
- Design assisted by testing
- Probability-based design

The partial safety factor method is widely used to design offshore structures. The partial safety factor method is based on a separate assessment of the load effects in the structure due to each applied load process. For example, consider the design of a transitional part of a jacket hybrid marine platform. Load combinations taking into consideration the linear summation of different loads with their load factors are considered. A load combination represents a load case, e.g. transportation or lifting.

For a specific load case, the loads are calculated, the load factors are applied, and then, the structure is subjected to this representative summation load. Alternatively, the structure is subjected to each load, and the load effects are multiplied to load factors. Afterward, the factored load effects are summed up and compared to the defined factored resistance (considering the material factors). Later in this section, the design based on partial safety factors is formulated.

Separate assessment of the load effects is accurate when the load effects as well as loads are independent. Otherwise, direct simulation of the combined load effects of simultaneously applied load processes is needed. This is demanding for offshore wind turbines and hybrid energy structures as the wave and wind loads are affecting each other. For instance, consider wave-induced aerodynamic actions (loads, damping, etc) or wind-induced hydrodynamic actions (refer to Karimirad 2011). The dynamic motions due to wave loads cause the turbine see an extra velocity and hence, produce more power. This means wave and wind actions are tightly connected, and the structure should be simultaneously subjected to both loads in a coupled integrated time-domain dynamic analysis.

Experiments, full-scale and model scaling of structures have been widely used to support design of marine structures. Such laboratories and trial tests can be used to determine load effects and resistance of structure. Test can be an alternative to analytical methods, however, in several cases as a supplement to analytical methods. Testing and observation of the structural performance of the models or full-scale structures are necessary for designing renewable offshore energy structures due to their complicated structural responses and dynamics under combined wave and wind loading. In the previous chapters, some examples of full-scale and model-scale testing of offshore wind turbines, wave energy converters and hybrid wave-wind concepts are mentioned.

Probability-based designs of marine structures are helpful for cases where limited experience exists, e.g. for offshore energy structures, especially for hybrid energy concepts. Structural reliability analysis methods are mainly considered for special design cases. Alternatively, they are used for calibrating the load and material factors to be used in the partial safety factor methods and to design special structures and innovative concepts.

7.9 Design Safety

Design safety has different aspects, such as safety class and target value of safety. Risk is a function of probability of failure and the consequences occur by a specific failure, for example:

$$\text{Risk} = \sum_{\text{allaccidents}} P(failure) \times Consequence. \tag{7.1}$$

This means for an acceptable risk level, we may allow higher failure frequency for components, which their failures have smaller consequences, e.g. no danger to

human. Standards classify structures into a safety class based on the failure consequences. The classification is normally determined by the purpose of the structure.
The purpose and functionality of the structure determines the classifications, in general.

For each safety class, a target safety level is defined in terms of a nominal annual
probability of failure. Det Norske Veritas (DNV 2013b Design of offshore wind
turbine structures) has defined three safety classes for offshore wind farms:

- Low-safety class: structures whose failures imply a low risk for personal injuries, pollution and economical consequences as well as negligible risk to human
  life.
- Normal-safety class: structures whose failures imply some risk for personal injuries, pollution or minor societal losses, or the possibility of significant economic
  consequences.
- High-safety class: structures whose failures imply large possibilities for personal
  injuries or fatalities, significant pollution or major societal losses, or very large
  economic consequences.

Support structures and foundations of renewable offshore energy structures, which
are normally unmanned, can be designed with normal-safety class. However, based
on economical motivations and considerations about human safety, other safety
classes may be applied. For example, consideration is needed for the maintenance
and consequent presence of human. Also, weather window/environmental conditions in connection with such cases should be reflected on the design. In addition,
to protect the investments in an offshore energy farm, design could be set to a high-
safety class. The safety class requirements are reflected in material or load factors,
e.g. DNV applies it to load factors for offshore wind and keep the material factors
unchanged regardless of safety class used in designing.

IEC61400-1 defines a normal-safety class for wind turbines. Considering the
normal-safety class for offshore wind turbines, DNV suggests that the target safety
level for support structures/foundation design have a nominal annual probability of
failure of 10^{-4} (this target safety is aimed for structures with ductile failures, e.g.
steel structures). The target safety level of 10^{-4} is compatible with the safety level
implied by DNV-OS-C101 for unmanned structures. For energy structures where
personnel present during harsh conditions, design to the high-safety class with a
nominal annual probability of failure of 10^{-5} is needed.

For an offshore energy farm (wave, wind or combined, including hybrid devices), proper risk assessments should be set up. This requires the consideration of
limit states especially ALS. Collision between ships and energy devices might occur. Hence, the development of a collision risk model and quantitative collision risk
assessments should be documented. Traffic data in the coastal area can help correct
and appropriate safety measurements for the farm. For floating units, mooring line
can fail during harsh conditions or due to fatigue. Drifting floats may hit the ships
in traffic or the other energy devices, and make damage and failure. Simulations
should verify the effects and assess the damage to correctly quantify the risk of
those situations.

7.10   Design Using Partial Safety Factor Method

Load and resistance factors are applied in the partial safety factor method. The design target safety is achieved by applying load and resistance factors to the characteristic values of the governing variables (loads, resistance, load effects and material strength). The design criterion is expressed in terms of the partial safety factors and characteristic values. The governing variables are:

- Loads acting on the structure or load effects in the structure
- The resistance of the structure or strength of the materials used in the structure

The characteristic values of loads and resistance (or of load effects and material strengths) are chosen as specific quantiles in their respective probability distributions. An example for governing variables used in "partial safety factor method" is given in the following part:

- Loads: environmental loads, e.g. wave, wind and ocean current loads
- Load effects: shear forces/bending moments at different sections of the structure
- Resistance: the capacity of the structure, e.g. critical values of deformations
- Material strength: yield strength of specific steel, e.g. 350 MPa

Designing of a structure or a structural component is safe when $S_d \leq R_d$, in which, S_d and R_d are design-load effect and design resistance, respectively. The design-load effect (S_d) is the combined load effect resulting from the simultaneous occurrence of loads:

$$S_d = fun(F_{d1}, F_{d2}, \ldots F_{dn}). \tag{7.2}$$

According to the partial safety factor format, S_d resulting from the occurrence of independent design loads can be written as:

$$S_d = \sum S_{di}(F_{ki}), \tag{7.3}$$

where $S_{di}(F_{ki})$ denotes a design-load effect (S_{di}) as a function of a specific characteristic load (F_{ki}).

There are two approaches to define the design-load effect (S_{di}):

1. S_{di} is obtained by the multiplication of the characteristic load effect (S_{ki}) by a specified load factor (γ_{fi}):

$$S_{di} = \gamma_{fi} S_{ki}, \tag{7.4}$$

where the characteristic load effect (S_{ki}) is determined in a structural analysis for the characteristic load (F_{ki}).

Table 7.1 Example[a] of load factors for ULS

ULS	Permanent loads	Variable loads	Environmental loads
Type A	1.3	1.3	0.7
Type B	1.0	1.0	1.3

[a] These values are concept/structure dependent and vary in standards, e.g. they may be different for an ocean oil platform compared to an offshore wind turbine

2. The design-load effect (S_{di}) is obtained by a structural analysis for the design load (F_{di}):

$$F_{di} = \gamma_{fi} F_{ki}, \tag{7.5}$$

where F_{ki} is characteristic load and γ_{fi} is a specified load factor.

The safety factors are defined in a way that the possible poor realizations of governing values and uncertainties are covered to ensure a satisfactory safety level. Load factors account for:

- Deviations of the loads from their characteristic values
- Several loads may simultaneously exceed their respective characteristic values
- Uncertainties in the numerical model and analysis

An example of load factors for ULS is mentioned in Table 7.1 (NORSOK 2004a Design of steel structures).

Characteristic values of load effect (S_{ki}) and load (F_{ki}) are obtained as specific quantiles in the distributions of the respective load effects (S_i) and respective loads (F_i), correspondingly. Quantiles selected may depend on which limit state is considered and vary from one specified combination of load effects to another. For example, for ULS analysis of fixed wind turbines, response/load with return period of 50 years is desired (for a floating wind turbine, this may be set to 100 years return period). The 50-year load or load effect corresponds to 98 % quantile in the distribution of the annual maximum of combined load or load effect.

If the representation of the dynamic responses is the main issue, the first approach should be used. If the representation of nonlinear material behaviour and geometrical nonlinearities are the main issue, the second approach is appropriate. An example of using these approaches is the determination of design load effects in structure using Approach 1 and design of foundation using Approach 2. For a monopile bottom-fixed wind turbine where the influences of soil nonlinearities are important, the following procedures can be used (DNV 2013b Design of offshore wind turbine structures):

a. First, an integrated structural analysis of the tower and support structure subjected to wave and wind loads plus permanent loads is performed using Approach (1) to find load effects at an interface (e.g. at the tower flange). The results of this analysis are shear forces combined with bending moments at that interface (the load effects).

b. Then, these design load effects (bending and shear forces) are applied as external loads at the chosen interface. And, the design load effects in the monopile structure and foundation pile for these design loads can then be determined from a structural analysis of the monopile structure and foundation pile considering the soil interactions by applying Approach (2).

For offshore energy structures subjected to simultaneous wave and wind loading, it may not be always feasible to determine the design load effect by a linear combination of separately determined individual load effects. For example, the total damping of a wind turbine depends on the wind loading and its direction relative to other loads. Hence, the structure should be analyzed for combined load effect for a simultaneous application of the wind and wave loads. In order to apply the partial safety factor method in such cases:

1. Structural analysis considering simultaneously applied load processes to define the distribution of combined load effect (S) is carried out.
2. The characteristic combined load effect (S_k) needs to be defined as a quantile in the upper tail of the distribution of the combined load effect.
3. Design combined load effect (S_d), resulting from the simultaneous occurrence of the loads, is established as a characteristic combined load effect multiplied by a common load factor:

$$S_d = \gamma_f S_k. \tag{7.6}$$

This highlights the importance of the integrated/coupled aero-hydro-servo-elastic and coupled time-domain analyses for offshore energy structures especially for hybrid devices and floating wind turbine concepts. Further, in this book, dynamic and structural responses covering the modelling and analyses aspects for offshore energy structures subjected to wave and wind loads are gradually discussed.

The resistance (R) against a particular load effect (S) is usually a function of parameters, such as geometry, material properties, environment and load effects themselves (e.g. through interaction effects such as degradation). Depending on the design situation, there are two approaches to establish the design resistance (R_d) of the structure or structural components (DNV 2013b Design of offshore wind turbine structures):

1. Design resistance (R_d) is obtained by dividing the characteristic resistance (R_k) by a material factor (γ_m):

$$R_d = \frac{R_k}{\gamma_m} \tag{7.7}$$

2. Design resistance (R_d) is obtained from the design material strength (σ_d) by a capacity analysis: $R_d = \text{fun}(\sigma_d)$ in which

$$\sigma_d = \frac{\sigma_k}{\gamma_m} \tag{7.8}$$

The characteristic resistance (R_k) and material strength (σ_k) are obtained as a specific quantile in the distribution of the resistance and material strength, respectively, which may be obtained by testing. Usually, the characteristic resistance (R_k) is defined as the 5 % quantile in the distribution of the resistance. Material factors account for:

- Deviations in the resistance of materials from the characteristic value
- Uncertainties in the model and analysis
- A possible lower characteristic resistance of the materials in the structure, as compared with the characteristic values interpreted from test specimens

Material factors depend on limit states, standards and materials (steel, aluminium, concrete, etc.). For example, NORSOK N-001, defined material factor of steel to be 1.15 for ULS analysis of offshore structures (NORSOK 2004b Norsok N-001, Structural design).

7.11   Design Using Direct Simulation of Combined Load Effects

Designs based on direct simulation of the combined load effects of simultaneously acting loads and the design using partial safety factor methods are similar. The main difference is that in the former case, the simulations are carried out by applying all the loads, simultaneously. This ensures accounting for the coupling and nonlinear effects between loads/load effects for the structures in which the linear combination of individual characteristic load effects determined separately for each of the applied loads is not valid.

For offshore energy structures (especially hybrid concepts and wind turbines), wave and wind loads are linked through the load effects. Wave-induced aerodynamics and wind-induced hydrodynamic loads and load effects, such as damping are tightly linked depending on environmental conditions and the status of the system. For example, the aerodynamic damping of a wind turbine depends on several issues:

- Presence of wave and wind
- Turbine operational conditions, it may be parked
- Misalignments between wind, wave and current

Also, the wave and wind are correlated and joint distribution of them results in dependency of wave and wind loads to each other. In these cases, hence, direct simulation of the characteristic combined load effects from the simultaneously applied loads should be implemented.

In the direct method, the loads are simultaneously applied and the characteristic load effects are obtained. Afterward, the design combined load effect (S_d) is obtained by the multiplication of the characteristic combined load effect (S_k) by a specified load factor (γ_f). The design format is similar to what introduced in partial safety method:

$$S_d = \gamma_f S_k. \tag{7.9}$$

For offshore energy structures, time-domain simulations considering both wave and wind loading with accounting for joint distribution of wave and wind are needed. In a structural time domain analysis of the turbine subjected concurrently to both wind and wave loading, the resulting combined load effect is correctly representing the coupling. The importance of integrated analyses increases floating concepts and for those involving both wave and wind devices in hybrid formats.

7.12 Design Certification of Wind Turbines

The aim of this section is to explain the main sequences and requirements for a MW wind turbine design under standards, rules and guidelines (Woebbeking 2007). Similar to other devices, wind turbines and wind farms need to get certification. According to the European standard EN 45020, certification is the confirmation of compliance of a product or a service with the defined requirements (e.g. guidelines, codes and standards).

In the field of wind energy, the focus lies on complete wind turbine components, such as rotor blades, gearboxes or towers. The scope consists of the examination of the structural integrity, safety and compliance with these requirements. The evaluation or assessment of the design is generally carried out in sequential steps:

- The first part covers all aspects of the safety and control concept as well as load assumptions and load calculations. The load calculations of wind turbine should be based on aero-elastic codes using stochastic wind fields and modal or finite element analysis techniques.
- During the second part of the design evaluation, all components (e.g. machinery, tower and electrical equipment) are examined on the basis of the previously approved loads considering the relevant standards and guidelines.

If the dynamic analysis of the system is not part of the general load calculations, it will be examined in parallel with the conformity assessment of the components. At the end of the design evaluation, manuals and procedures for manufacturing, transport, erection, start-up, commissioning, operation and maintenance are checked for suitability, completeness and compliance with the assumptions in the design documentation. In addition to this, the evaluation of personnel safety is important. Rotor blade testing forms an integral part of the Type Testing of the blade. Lightning protection will be assessed in combination with the electrical equipment. A flowchart of the design evaluation is shown in Fig. 7.4.

In order to certify a wind park, project certification is needed. It is obvious that prior to the project certification, the type certification has to be issued. A type certi-

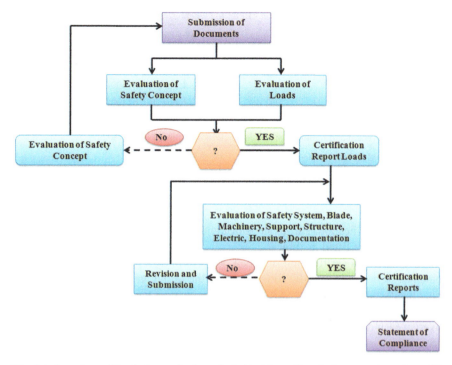

Fig. 7.4 Procedure of the design evaluation of wind turbines. (For similar examples refer to GL 2010; Woebbeking 2007; DNV.GL 2012)

fication is a certification for a wind turbine, which is the basis of a project certification. The project certification issued for a wind farm includes several wind turbines of same or different types. According to the IEC, in general, type certification comprises:

- Design evaluation
- Manufacturing evaluation
- Type testing
- Foundation design evaluation
- Type characteristics measurement

Within the design evaluation, some items like evaluation of manufacturing plan as well as installation plan and evaluation of personnel safety are necessary for IEC. The prototype testing includes an evaluation of a dynamic blade testing and the manufacturing evaluation includes an evaluation of the quality system as well as the inspection of the manufacturing.

Project certification on basis of the IEC is based on type certification and covers the aspects of site assessment, foundation design evaluation and installation evaluation. These individual modules are concluded with conformity statements. Certificates are issued upon the successful completion of the relevant type certification

Table 7.2 Example[a] of design evaluation. (Refer to (Woebbeking 2007)

Steps of evaluation	Codes or standards to be applied
Load assumptions	GL-Guideline; IEC61400-1 2nd or 3rd edition
Safety system and manuals	GL-Guideline; IEC61400-1 2nd or 3rd edition
Rotor blades	GL-Guideline; IEC61400-23
Machinery components	GL-Guideline; IEC61400-1 3rd edition
Tower (and foundation)	GL-Guideline
Electrical equipment and lightning protection	GL-Guideline; IEC61400-24
Nacelle housing and spinning	GL-Guideline

[a] These are just examples, which can vary in different standards and may be changed over time in new revisions of IEC codes, GL, DNV and DNV.GL standards

and project certification. Partial steps of the design evaluation and relevant codes and standards are listed in Table 7.2.

To get the full certification of a new wind turbine, a prototype of a wind turbine has to be made in order to evaluate all the design calculations with the real time measurements. Type certifications on the basis of the GL are divided to four consequent levels: D-Type, C-Type, B-Type and A-Type. These four certifications overlap each other. As a minimum, to certify a new design, D-design assessment is required.

I. D-type and C-type Design Assessment and Certification of Wind Turbines Within the D-design assessment for prototypes of wind turbines, a plausibility check of the prototype will be performed on the basis of the design documentation (DNV.GL 2012), (GL 2010) and (Woebbeking 2007). It is based on a load assessment and a complete check of the rotor blades and the machinery without the information about the site. This certification shows the approval of the overall wind turbine system. A list of the documents that are needed for this step are as follows:

- General description of wind turbine, including energy conversion concept (e.g. generator-conversion system)
- List of the primary components to be used (main bearing, gearbox, break, generator, converter, etc.)
- Description of the control and safety system
- Description of the electrical installation, at least inside the hub, nacelle and tower
- Concepts of the lightning protection system
- Results of complete load assumptions
- Main drawings of the rotor blades
- General arrangement drawings of the nacelle
- Drawing of the hub, main shaft, and main frame
- Main drawing of the gearbox
- Data sheet of the main electrical components
- Main drawings of the tower
- Calculation documentation for load assumptions

The C-design assessment is similar to D-design assessment with the additional information about the foundation and the owner of the prototype (DNV.GL 2012), (GL 2010) and (Woebbeking 2007). It means if the owner has a plan to make a prototype after the design evaluation, he can directly apply for C-type certification, and C-type certification is enough to make a prototype. For the C-design assessment, the following documents in addition to the documents have been mentioned for D-design assessment is required:

- Main drawing of the foundation
- Soil investigation report
- Name and address of the owner
- Planned location of the prototype
- 10-min mean of the extreme wind speed at hub height with the recurrence period of 50 years and the mean air density for the planned location of the prototype
- Test report for the arc resistance test of medium-voltage switchgear

This type of design assessment can be used to erect the prototype of a wind turbine as well as of the tower and foundation. Depending on national or local regulations the complete assessment of the tower and foundation might be necessary. The final step of this assessment will be the issue of a Statement of Compliance for the C-design assessment, which is valid for test operation comprising a maximum of 2 years or 4000 equivalent hours at full load. Now, the C-type certification is issued and the owner can apply for A- and B-design assessment.

I. B-type and A-type Design Assessment and Certification of Wind Turbines To attain the A- or B-design assessment, a complete examination of the design analysis with all required materials and component tests is required together with witnessing of the commissioning of one of the first wind turbine (DNV.GL 2012), (GL 2010) and (Woebbeking 2007). The consequence of the procedure is illustrated in Fig. 7.5.

The difference between the A- and B-design assessments is only in the outstanding items. The B-design assessment may contain items that are still outstanding, providing these are not directly relevant to safety. The final (A-design) assessment is issued if there are no outstanding items. The statement of compliance for the A-design assessment is valid indefinitely if no modification is made (DNV.GL 2012), (GL 2010) and (Woebbeking 2007).

For the assessment of the design documentation, the manufacturer shall submit a full set of documents in the form of specifications, calculations, drawings, descriptions and part lists. The documents, which form the basis of the design assessment are:

- Control and safety system concepts
- Load case definitions and assumptions for

  - Rotor blades
  - Mechanical structure
  - Machinery components
  - Electrical installation
  - Tower and foundation

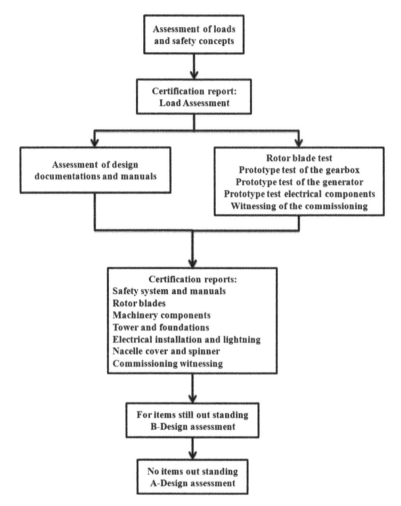

Fig. 7.5 Procedure of assessments of wind turbines. (Refer to GL 2010 and DNV.GL 2012)

- Manuals

  - Electrical
  - Commissioning
  - Operation
  - Maintenance

In addition to these certifications, during the erection of the wind turbine prototype, all the requirements mentioned in the design documentations are examined and quality management system for manufactures shall be inspected.

7.13   Design Loads for Offshore Wind Turbines

This section gives an introduction to the requirements for load components, load combinations and load cases to be considered for designing offshore wind turbines. Reference is made to IEC (International Electrotechnical Commission), DNV (Det norske Veritas) and GL (Germanischer Lloyd) e.g. IEC61400-1, IEC61400-3, DNV-OS-J101, DNV-OS-J103 and (GL 2010). The main point is to design an offshore wind turbine, which withstands all the possible loads and load combinations during its life. A wind turbine meets different conditions and situations including temporary and permanent design status:

- Temporary design status includes conditions during transport, assembly, maintenance, repair and decommissioning of the wind turbine structure.
- Permanent phase includes both steady states and transient cases. Steady conditions cover power production, idling and parked (stand-still) while transient conditions are associated with start-up, shutdown, yawing and faults.

The characteristic load and its role in design format are explained in the partial safety method. For temporary cases, the location and time are highly important: such as transportation and installation with special requirements needed for weather window required for those operations. For the temporary design conditions, the characteristic loads and load effects in design checks are selected either based on specified environmental design conditions or based on specified design criteria. The design criteria shall be specified with due attention to the actual location, the season of the year, the weather forecast and the consequences of failure. Also, the design criteria shall be selected for all temporary phases and will depend on the measures taken to achieve the required safety level.

For permanent cases including operational conditions, the characteristic loads are selected depending on load nature and limit states. Standards, rules and regulations, such as DNV, GL and IEC, have recommendations and requirements for the selection of characteristic loads. Here, some examples are given below to highlight this issue (DNV 2013b Design of offshore wind turbine structures).

ULS: permanent (expected value), variable (specified value), environmental (load/load effects with return period of 50 years), abnormal turbine loads (specified value), deformation (expected extreme value). The texts in brackets define how the characteristic value of a load should be chosen. For instance, the environmental loads (wind, wave and current loads) for a bottom-fixed wind turbine should have return period of 50 years. This is equivalent with 98 % quantile in the distribution of annual maximum of the loads.

FLS: All loads have the same characteristic values as for ULS except for environmental loads which should have expected load history (or expected load effect history).

For more information regarding characteristic load selection of different loads for a specific limit state depending on the structure type (land-based, bottom-fixed offshore and floating wind turbines) refer to standards and regulations mentioned in this section. Now, the primary loads important for the design of offshore energy structures (mainly offshore wind turbines) are discussed (DNV 2013a Design of floating wind turbine structures).

1. Permanent loads: these loads are mainly gravity and hydrostatic pressure loads, which do not vary in magnitude, position or direction during the period considered[1]. Examples are mass of structure, mass of permanent ballast and equipment, and external and internal hydrostatic pressure. The characteristic value of a permanent load is defined as the expected value based on accurate data of the unit, mass of the material and the volume in question.

2. Variable functional loads: loads, which may vary in magnitude, position and direction during the period under consideration. Examples are: personnel, crane operational loads, ship impacts, loads from fendering, loads associated with installation operations, loads from variable ballast and equipment (as well as stored materials, gas, fluids and fluid pressure and lifeboats). For an offshore wind turbine structure, the variable functional loads usually consist of:

   a. Actuation loads: operation and control of the turbine result in actuation loads, such as torque control from a generator (or inverter), yaw and pitch actuator loads and mechanical braking loads. Actuation loads are usually accounted for through aerodynamic load calculation of wind turbine subjected to environmental wind loading.
   Actuator loads should be accounted for in the calculation of loading and response, e.g. for mechanical brakes: the range of friction, spring force or pressure as influenced by temperature and ageing should be considered in load-effect analyses during braking events.

   b. Loads on access platforms and internal structures: these loads are relevant for local design; therefore, do not appear for designing primary structures and foundations.

   c. Ship impacts from service vessels: these loads are used for designing primary support structures and foundations and for some secondary structures. DNV offshore standards urge that the impacts from approaching ships in the ULS shall be considered as variable functional loads while impacts from drifting ships in the ALS shall be considered as accidental loads.

   d. Crane operational loads

The characteristic value of a variable load is the specified value that produces the most unfavourable load effects in the structure.

3. Environmental loads: these loads are caused by environmental phenomena, such as wave, wind, current and tides. There are other parameters, which may be important for specific marine structures depending on the site conditions and

---

[1] For floating structures: magnitude, position or direction of these loads may change due to the motion of the structure. However, inherently, the load is not changing when the floater is not moving. In some cases, a permanent load for a fixed structure may be considered variable for a floating structure.

concept. These parameters are ice, earthquake, soil conditions, temperature, fouling and visibility; practical information regarding environmental loads and environmental conditions is given in DNV-RPC205 (DNV 2007, Environmental conditions and environmental loads) . These loads may vary in magnitude, position and direction during the period under consideration, and which are related to operations and normal the uses of the marine structure are as follows:

- Wind and aerodynamic loads
- Wave and hydrodynamic loads; loads induced by waves and current, including drag forces and inertia forces, diffraction and radiation effects, etc.
- Seismic and earthquake loads
- Ocean current loads
- Tidal effects
- Marine growth and fouling
- Snow, rain and ice loads

Characteristic environmental loads and load effects are usually determined as quantiles with specified probabilities of exceedance. The environmental loads mostly have stochastic nature. The sensitivity analyses considering statistical methods are carried out for measured data or simulated data. These statistical analyses are performed both for metocean data, loads and load effects to study the sensitivity and uncertainty caused by stochastic processes. The validation of applied methods and distributions considering data processing are tested. The longest possible time-period for the relevant offshore site should be used. If short time series are implemented, statistical uncertainty shall be accounted in the determination of characteristic values.

Due to the importance of aerodynamics (including wind loads) and hydrodynamics (including wave loads) for offshore energy structures, these issues are separately discussed in the following chapters of this book. The other relevant environmental loads are briefly explained in this section.

- Ice loads
  Laterally, moving ice causes loads on offshore support structures. These loads are usually difficult to be assessed. They depend on the nature and properties of the ice, including the age of the ice, the salinity and temperature of the ice. The strength of the ice decreases for aged and saline ice. Hence, the younger and less saline ice has larger forces on the structure. The prediction of the ice loads has uncertainties involved, and care is needed when accounting the ice loads. Ice loads are thoroughly explained in ISO 19906 (ISO 2010). When dealing with moving ice, the following issues are important:

  - Magnitude and direction of ice loads (ice movement)
  - Frequency of the ice loads
  - Nature of the ice
  - Mechanical properties of the ice
  - Ice-structure contact area
  - Size and shape of the structure

- Oscillating nature of the ice loads, including build-up and fracture of moving ice
- Full-scale measurements, model experiments (which can be reliably scaled) and recognized theoretical methods can be applied to calculate the loads

Different design guidelines and standards with emphasize on the dominant sea ice loads for offshore support structures for wind turbines in the subarctic exist. The documents published by Germanischer Lloyd (GL), International Electrotechnical Commission (IEC), Det Norske Veritas (DNV) and The International Organization for Standardization (ISO) are the key standards and guidelines for OWTs.

Often, these standards and guidelines refer to offshore oil and gas technology for calculation of the ice loads which may not be accurate for OWTs application (in some cases). Research is going on to highlight the requirements needed for OWTs and marine energy applications. Popko et al. (2012) have studied the difference of the ice load calculation in different standards and guidelines in a paper called "State-of-the-art Comparison of Standards in Terms of Dominant Sea Ice Loads for Offshore Wind Turbine Support Structures in the Baltic Sea".

- Water level loads
  When considering the responses and load effects, correct water level considering the storm surge and tide should be evaluated. Hence, tide and storm surge effects should be considered in the evaluation of the loads. Usually, by increasing the water level due to storm surge and tide, the hydrostatic loads and current loads on the structure increase. However, situations may exist where lower water levels will imply the larger hydrodynamic loads. Also, the air gap decreases for higher mean water levels. Figure 7.6 shows the water level definitions; a

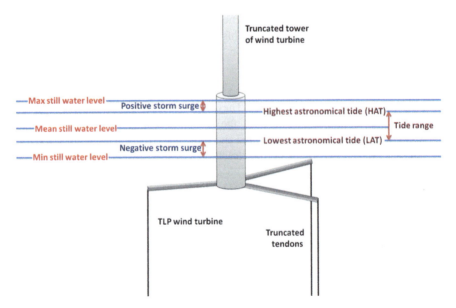

Fig. 7.6 Water level definition

schematic layout of a TLP wind turbine is plotted with respect to different water levels. Due to increase of water level, the loads on the platform increase and tension in tendons gets higher. Care is needed when considering and evaluating the breaking tension in tendons. When the water goes down, the tension decreases which in combination to wave and relative motion of platform can cause slack. The zero tension in the lines and cyclic loading can cause fatigue of the lines as well as excessive stress in the local connections to arms (pontoon).

- Seismic and earthquake loads
  When a marine structure, e.g. offshore energy structure, is designed for an offshore site, which may be subjected to an earthquake, the earthquake loads should be considered. Response spectra in terms of so-called "pseudo" response spectra may be used for this purpose. If the structure is in areas which are subjected to tsunami set-up by earthquakes, the load effects of the tsunamis on the structure should be considered as well. In earthquake analyses, ground motions are either considered as response spectra or time histories (NORSOK 2004a Design of steel structures). For cases when soil–structure interaction is significant and for deep-water bottom-fixed piled-structures (e.g. jackets), time-history analysis is suggested. General requirements concerning earthquake analyses are given in standards like NORSOK, "Actions and Actions Effects", N-003.

- Marine growth and fouling
  Marine growth increases the hydrodynamic and gravity actions. Fouling and marine growth must be taken into account by increasing the outer diameter of the structural member when calculating the hydrodynamic wave and current loads. When considering marine growth, the following issues are important (see Fig. 7.7):

  - The thickness of the marine growth depends on the depth below the sea level and the orientation of the structural component.
  - The thickness may be assessed based on relevant local experience and existing measurements.

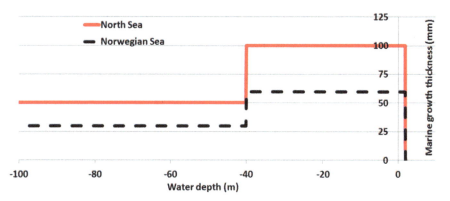

Fig. 7.7 Example of marine growth thickness variation with respect to water depth (DNV 2013b Design of Offshore Wind Turbine Structures)

- Site-specific studies help to establish the likely thickness and depth dependence of the growth.

When specific information at the offshore site is not available, the recommended marine growth thickness from standards and regulations should be chosen. For different ocean regions in the world, the amount of fouling (micro and macro biofouling) is dependent on water depth, water parameters, offshore site location and its characteristics, etc.

Usually, a plan for inspection and fouling removal is considered as part of the marine structural designing, as there are uncertainties involved in the assumptions for the marine growth (biofouling) on structures. Based on the impact of fouling on the performance and structural integrity (safety and reliability), such a plan is set considering the inspection frequency, applied methods and the criteria of fouling removal. This kind of plan is usually based on experiences with marine-growth under consideration for specific conditions.

- Scour
  Due to water kinematics at seabed as well as wave and current loads (depending on depth, offshore site location and soil characteristics), soil particles move. This erosion of soil around the piles and near the structure foundation is so-called scour. Scour may have an impact on the geotechnical capacity of a foundation. Hence, the structural responses that govern the limit states of components, such as ULS and FLS, could be affected. Means to prevent scour and requirements to such applications should be considered in designing. Bottom-fixed offshore energy structures, such as monopile wind turbines, are mostly affected by scour.

- Design loads due to marine operations
  Specific attention is needed for marine operations, for transportation, installation and maintenance of offshore energy structures. This mainly includes loads during load-out, transportation and installation. The sub-structure may be transported afloat, e.g. wet-towed (such as the Hywind case) or the commissioned system can be transported afloat like WindFloat (refer to previous sections for more information about these offshore energy structures). Barge may be used to transport the structures; and crane barge may be used in order to lift or upright the structure for installation depending on the type of structure and the planned transportation/installation methods.

  Acceptable environmental conditions during marine operations (including transportation and installation of marine structures and their foundations) based on safety aspects and design criteria should be defined. This is so-called "weather window" in which a specific marine operation depending on the response of the structure is allowed. During installation, dismantling and replacement of wind turbine components, such as rotors and nacelles, the design criteria are defined for acceptable environmental conditions. These temporary phases influence the structures and foundations (their safety and structural integrity), e.g. by exposing them to temporary loads. When considering the acceptable environmental conditions and weather window, the following items should be considered (DNV 2013a Design of floating wind turbine structures):

- – Wind speed and wind direction
- – Significant wave height and wave direction
- – Wave height and wave period
- – Water level (tide)
- – Current speed and current direction
- – Ice

Detailed information of wave and wind (metocean) are needed for marine operations. Not only the significant wave height but also the heights of individual waves (based on an on-site assessment) are usually needed. The significant wave height is used in the marine operation calculations during the operation itself, and the individual wave heights are needed for accurate actions. Water level may be an issue in locations with significant tide; and, ice may be an issue for maintenance and repair operations in harsh climates (for more information regarding marine operation refer to DNV 2011 Modelling and analysis of marine operations).

7.14   Design Load Cases

To design marine structures, different load cases based on relevant design loads and their combinations are considered. This considers the status of the structure, its functioning, environmental loads and similar aspects. For offshore energy structures (WECs, OWTs and Hybrid marine platforms), depending on the status of the power take off or wind turbine, more detailed load cases are needed to cover all possible scenarios during the life of the structure. Different design situations may govern the designs of different components of the system, support structure and the foundation.

For designing the structure and foundation, several load cases corresponding to different design situations of the system are considered; among those are wind turbine loads due to wind load on the rotor and on the tower as well as wave loads on the wave energy converter.

The load cases should be defined such that the structural integrity and safety of the structure in its design life are ensured. All the limit states should be covered by the defined load cases. The 50-year or 100-year loads/load effects (depending on applied standards and regulations) are considered for each structural part in the ULS analysis. Also, the load cases are defined to guarantee that they capture all contributions to fatigue damage (considering design requirements for FLS analysis). The load cases capturing abnormal cases and associated with fault conditions should be investigated as well.

Offshore energy structures (especially hybrid marine energy devices and OWTs) are simultaneously subjected to wind, water level, current and wave loads. Hence, the load cases to be considered must specify not only the wind turbine load conditions, but also their accompanying wave, current, water level and similar environmental conditions (e.g. ice if applicable and similar issues). Most importantly, an appropriate combination of wind and wave loading is necessary for designing purpose in an integrated analysis.

The IEC (International Electrotechnical Commission) has issued the 61400-3 standard, which describes 35 different load cases for design analysis of offshore wind turbine. In the IEC standard, different load cases are introduced for offshore wind turbines to ensure the integrity of the structure in installation, operation and survival conditions (Karimirad 2011). The defined load cases are given below:

a. Power production
b. Power production plus fault condition
c. Start-up
d. Normal shutdown
e. Emergency shutdown
f. Parked
g. Parked plus fault condition
h. Transport, assembly, maintenance and repair

The power production load case is the normal operational status in which the turbine is running and is connected to an electrical load with active control. This means the turbine generates electricity in regular conditions.

If the fault occurs and turbine continues to produce electricity, the power production plus fault condition happens. This involves a transient event triggered by a fault or loss of electrical network connection while the turbine is operating under normal environmental conditions. If this case does not cause immediate shutdown, the resulting loads could affect the fatigue life, which should be analyzed while checking the FLS.

Start-up is a transient load case. The number of occurrences of start-up may be obtained based on the control algorithm behaviour. This transient load case can affect the fatigue life of components and should be included in FLS.

Normal shutdown and emergency shutdown are transient load cases in which the turbine stops producing power by setting the parked condition. The rotor of a parked wind turbine is either in the standstill or idling condition. The ultimate loads for these conditions should be investigated. In some cases, survival conditions may govern the ULS cases, especially for floating wind turbines. Any deviation from the normal behaviour of a parked wind turbine, resulting in a fault should be analyzed. When combining the fault and extreme environmental conditions in the lifetime, a realistic situation should be proposed. Defining the probability of occurrence of fault conditions in survival environmental cases is challenging (If such fault happens, it can destroy part of the structure). An example can be a floating hybrid marine platform under survival wave-wind loads while having a fault in either its control system or actuators (servo).

All the marine conditions, wind conditions and design situations should be defined for the transport, maintenance and assembly of an offshore wind turbine. The maximum loading of these conditions and their effects should be investigated (refer to previous sections of this chapter regarding the design loads for marine operations).

Fatigue and extreme loads should be assessed with reference to material strength, deflections and structure stability. In some cases, it can be assumed that the wind and waves act from one direction (single-directionality). In some concepts, multi-directionality of the waves and wind can be important. In the load case with transient change in the wind direction, it is suggested that co-directional wind and wave be assumed prior to the transient change. For each mean wind speed, a single value for the significant wave height (e.g. expected value) can be used. This method has been used to analyze the wave- and wind-induced response of offshore wind turbines, e.g. based on IEC standard.

The number of load cases varies in different standards; for offshore wind turbines, there are 30–40 load cases. The point is to cover all possible load combinations that a marine energy structure, e.g. FWT is phasing in its design life. Here, some examples of the load cases are discussed based on DNV, GL and IEC standards. Table 7.3 specifies examples of design load cases to be considered for offshore wind turbine load conditions; the wind load conditions joint with wave load conditions, current and water level conditions are listed in this table. The load cases refer to design for ULS and FLS. The table includes a number of abnormal load cases for the ULS analysis. For offshore wind applications, the load cases are usually defined in terms of wind conditions which are characterized by wind speed. For most of the load cases, the wind speed is defined as a particular 10-min mean wind speed plus a particular extreme coherent gust, which forms a perturbation on the mean wind speed. Some load cases refer to normal wind profiles. For each specified load cases, simulations for simultaneously acting wind and waves are carried out.

a. $H_{s,1}$ h = $\beta H_{s,3}$ h: in which, the combined conversion and inflation factor (β) is 1.09 if wave heights are Rayleigh-distributed and the number of waves in 3 h is around 1000
b. If the simulation period is longer than the averaging period for the mean wind speed, a deflation factor may be applied. For the simulation period of 1 h and the averaging period of 10-min, the deflation factor may be taken as 0.95. (Refer to DNV 2013b Design of offshore wind turbine structures)

7.15 Design of Floating Wind Turbines

Design aspects for a floating wind turbine can be quite different in a sense that a floating structure is engineered instead of a fixed structure. However, there are common practices in designing floating structures and bottom-fixed marine structures. Basically, the bottom-fixed marine structures are being dealt in coastal engineering while floating ocean structures are mainly developed in offshore oil and gas technologies. The story of offshore wind technology is a bit different and multiple competencies covering coastal engineering, marine technologies, offshore industry

Table 7.3 Examples of design-load cases. (Refer to DNV, IEC and GL for more information)

Design situation, (DLC)	Wind condition ($V_{10,hub}$ or V_{hub})	Wave condition (H_s or H)	Wind and wave directionality	Current	Water level	Limit state / Partial safety factors
Power production, (1.2)	NTM $V_{in} < V_{10,hub} < V_{out}$	NSS H_s ($V_{10,hub}$) from joint probability distribution of H_s, T_p and $V_{10,hub}$	Co-directional in one direction	Note[a]	Between upper and lower 1-year water level	FLS, Note[b]
						Note[c]
Power production plus	NTM $V_{in} < V_{10,hub} < V_{out}$	NSS $H_s = E[H_s|V_{10,hub}]$	Co-directional in one direction	Wind-generated current	MWL	ULS, Control system fault or loss of electrical connection
						Normal
Start-up,(3.2)	EOG $V_{10,hub} = V_{in}$, V_{out} and $V_r \pm 2$ m/s	NSS $H_s = E[H_s|V_{10,hub}]$	Co-directional in one direction	Wind-generated current	MWL	ULS
						Normal
Normal shutdown, (4.1)	NWP $V_{in} < V_{10,hub} < V_{out}$ + normal wind profile to find average vertical wind shear across the rotor	NSS $H_s = E[H_s|V_{10,hub}]$	Co-directional in one direction	Note[a]	Between upper and lower 1-year water level	FLS
						Note[c]
Emergency shutdown, (5.1)	NTM $V_{10,hub} = V_{out}$ and $V_r \pm 2$ m/s	NSS $H_s = E[H_s|V_{10,hub}]$	Co-directional in one direction	Wind-generated current	MWL	ULS
						Normal
Parked, standing still or idling, (6.3a)	EWM Turbulent wind $V_{10,hub} = V_{10,1-yr}$ (standard deviation of wind speed: $0.11\,V_{10,hub}$)	ESS $H_s = H_{s,1-yr}$ Note[d]	Misaligned Multiple directions	1-year current	1-year water level	ULS, Extreme yaw misalignment

Table 7.3 (continued)

Design situation (DLC)	Wind condition ($V_{10,hub}$ or V_{hub})	Wave condition (H_s or H)	Wind and wave directionality	Current	Water level	Limit state	
						Partial safety factors	Extreme
Parked plus fault conditions, (7.2)	NTM $V_{10,hub} < 0.7 \cdot V_{10,50\text{-}yr}$	NSS $H_s(V_{10,hub})$ from joint probability distribution of H_S, T_p and $V_{10,hub}$	Co-directional in multiple direction	Note[a]	Between upper and lower 1-year water level	FLS	
						Note[c]	
Transport, assembly, maintenance and repair, (8.3)	NTM $V_{10,hub} < 0.7 \cdot V_{10,50\text{-}yr}$	NSS $H_s(V_{10,hub})$ from joint probability distribution of H_S, T_p and $V_{10,hub}$	Co-directional in multiple direction	Note[a]	Between upper and lower 1-year water level	FLS	
						Note[c]	

DLC design load case, the number here refers to DNV standard numbering (it may be different for some load cases in IEC or GL standards), $V_{10,hub}$ 10-min averaged wind speed at hub height, V_{hub} wind speed at hub height, *NTM* normal turbulence model; wind has a stochastic nature. The turbulence of the wind is represented by energy carried along the turbulence eddies. Distribution of wind energy over frequencies can be represented by power spectra and coherence functions. This representation is generally (at least for bottom-fixed wind turbines) considered as an adequate for turbulence over a period of approximately 10 min. The characterization of the natural turbulence of the wind (assuming statistic parameters for the relatively short period in condition that the spectrum remains unchanged), results in: (1) mean value of the wind speed, (2) turbulence intensity and (3) integral length scales, V_{in} cut-in wind speed in which turbine starts to operate, V_{out} cut-out wind speed in which turbine shutdowns, H_s same as wind, wave is stochastic; significant wave height is used to represent a sea state condition (together with other parameters), *H* individual wave height (it can be an extreme regular design wave, e.g. 100-year design value), *NSS* normal sea state, a sea state which is characterized by a significant wave height, a peak period (T_p) and a wave direction. It is associated with a concurrent mean wind speed, *MWL* mean water level (refer to Fig. 7.6), *EOG* extreme operating gust. Depending to wind turbine class, the extreme operating gust is defined by a formula in which the gust wind speed at the hub height is related to rotor diameter, return period (e.g. 1-year or 50-year) and some other parameters (like turbulence scale parameter), V_r rated wind speed, *NWP* normal wind profile model denotes the average wind speed as a function of height above the still water line and can be calculated by power law, *EWM* extreme wind speed model, EWM is usually based on investigations at the installation site; otherwise, it is defined in the standards, $V_{10,1\text{-}yr}$ 10-min averaged wind speed with recurrence of 1 year, $V_{10,50\text{-}yr}$ 10-min averaged wind speed with recurrence of 50 year, *ESS* extreme sea state is defined by a significant wave height, a peak period and a wave direction. The extreme sea state can have different return periods, e.g. 100 years. The range of peak periods appropriate to each of these significant wave heights should be considered and calculation with peak period values result in the highest loads and load effects in the structure need to be applied

Table 7.3 (continued)

Design situation, (DLC)	Wind condition ($V_{10,hub}$ or V_{hub})	Wave condition (H_S or H)	Wind and wave directionality	Current	Water level	Limit state
						Partial safety factors

[a] In general, ocean current acting simultaneously with wave and wind should be included, as the current influences the hydrodynamic loads and thus the fatigue damage (relative to the case without current). However, in some cases current effects can be ignored. Examples are when the wave loading is inertia-dominated or when the current speed is small. In these examples, the current loading is relatively negligible compared to wave loading

[b] The significant wave height (HS) of the normal sea state (NSS) is defined as the expected value of the significant wave height conditioned on the concurrent 10-min averaged wind speed. NSS is used for the calculation of ULS and FLS. For fatigue load calculations, a series of normal sea states have to be considered. Sea states are associated with different mean wind speeds; this is coming from the joint distribution of wave and wind. The number and resolution of these normal sea states should be checked to ensure accurate prediction of the fatigue damage associated with the full long-term distribution of metocean parameters. The range of peak periods (TP) appropriate to each significant wave height is considered (peak periods which result in the highest loads or load effects in the structure should be selected). For more information refer to the mentioned standards

[c] Partial safety factor for fatigue strength is usually considered equal to 1 for all normal and abnormal design situations. Some standards may allow lower load factors, e.g. GL allows smaller partial safety factors after consultation with GL Wind, if the loads were determined from measurements (or from analyses verified by measurements) with a higher level of confidence

[d] If simulation period (e.g. 1 h simulation) is shorter than the reference period of Hs (e.g. Hs,3 h), the significant wave height needs to be converted to a reference period equal to the simulation period (e.g. to have Hs,1 h). An inflation factor on the significant wave height needs to be applied in order to make sure that the shorter simulation period captures the maximum wave height, correctly

as well as wind technology are deployed. This includes a wide range of knowledge at least comprising of structural mechanics, fluid dynamics and control engineering. In this section, special issues related to the designing of floating wind turbines (FWTs) are mentioned. The core of the design is based on what is already discussed in this chapter. Now, specific topics for FWTs and similar aspects for other floating renewable offshore energy structures (e.g. hybrid marine platforms) are explained.

Design of wind turbine (land-based) and the related components of wind turbine, such as rotor, nacelle, generator and gearbox are well-established in standards, such as the IEC 61400-1. Design of bottom-fixed offshore wind turbines are also developed to a large extent; existing standards, regulation and recommendations from DNV, GL and IEC highlight this fact; e.g. DNV-OS-J101 covers design of fixed offshore wind turbines. Recently, floating hybrid energy platforms and FWTs are developed more; for example, DNV-OS-J103 attempts to provide requirements for designing floating wind turbines and related structural parts. This is mainly focusing on the structural part of the support structure, tower and mooring system for station keeping. The author kept an eye on the recent DNV-OS-J103 standard to bring in line the information.

For fixed-wind turbines, the tower is more or less connected to a rigid foundation/ support (similar to the connection of a land-based wind turbine to its foundation). However, a tower of a floating wind turbine has more flexibility at connection to the floater (support structure) and thus different overall stiffness due to the boundary conditions' effects. The stiffness of the tower forms the basis of the approval for a type-approved tower. The boundary conditions changed due to the flexibility of the connection between the tower-support interfaces will affect the stiffness and hence, endanger the assumptions used for type-approval. Hence, safety requirements assumed in type-approval need to be checked for a turbine mounted on a floater support. Changes applied on the tower characteristics such as structural damping, mass and stiffness as well as boundary conditions at the tower bottom interface alters the eigen-frequencies. Tower eigen-modes can couple with load and load-effects of rotor and increase vibrations, fatigue damage and in worst case produce instabilities resulting in failure modes. Type-approval of a land-based wind turbine is not necessarily guarantee a safe design for an offshore application. Time domain dynamic analyses considering these issues and accounting for modifications applied on the tower are highly recommended.

Floating wind turbines and floating hybrid marine platforms have usually low frequency responses. To capture these effects when loads and load effects are calculated, sufficient simulation length is needed. For land-based wind turbine, 10-min simulations are widely used and found to be adequate to present the corresponding wind load effects. For bottom-fixed offshore wind turbines, the simulation length should be increased, e.g. 30 min. For floating wind turbines, similar to other ocean structures, 3-h analyses are needed to capture nonlinearities, second-order effects, slowly varying responses, and to properly establish the design load effects. This is challenging as the wind is not stationary over long time scales, e.g. 3-h. One solution is to run three 1-h stochastic independent simulations and combine the results.

This should properly capture load and load effects for floating wind turbines (refer to Karimirad and Moan publications for more information).

References

DNV. (2007). *Environmental conditions and environmental loads*. Norway: Recommended Practice, DNV-RP–C205.

DNV. (2011). *Modelling and analysis of marine operations*. Norway: DNV-RP-H103.

DNV.GL. (2012). *Guideline for the certification of offshore wind turbines, edition 2012*. Hamburg: DNV.GL.

DNV. (2013a). *Design of floating wind turbine structures*. Norway: DNV-OS-J103, Det Norske Veritas.

DNV. (2013b). *Design of offshore wind turbine structures*. Norway: Det Norske Veritas AS January 2013.

GL. (2010). *Guideline for the certification of wind turbines*. Hamburg: Germanischer Lloyd.

ISO. (2010). *ISO 19906: Petroleum and natural gas industries—Arctic offshore structures*. International Organization for Standardization.

Karimirad, M. (2011). *Stochastic Dynamic Response Analysis of Spar-Type Wind Turbines with Catenary or Taut Mooring Systems*. PhD thesis, NTNU, Norway.

NORSOK. (2004a). *Design of steel structures. Lysaker*, Norway. www.standard.no/petroleum.

NORSOK. (2004b). *Norsok N-001, Structural design,* Norway. www.standard.no/petroleum.

Popko, W., Heinonen, J., Hetmanczyk, S., Vorpahl, F. (2012). State-of-the-art comparison of standards in terms of Dominant Sea ice loads for offshore wind turbine. *Proceedings of the twenty-second (2012) international offshore and polar engineering conference*. pp. ISBN 978-1 880653 94 1 (Set); ISSN 1098 6189 (Set).

Woebbeking, M. (2007). *Development of a new standard and innovations in certification of wind turbines*. Germany: GL.

Chapter 8
Wave and Wind Theories

8.1 Introduction

As it is mentioned in the previous chapter, wave and wind are the main sources of environmental loads. The first step in performing rational structural dynamic analysis to find load-effects is setting realistic environmental conditions. The most important for renewable offshore energy structures are the wind and wave at the park site. However, at some offshore locations, other parameters may be important (e.g. air and sea temperature, tidal conditions, current and ice conditions); some of those are mentioned in the previous chapter.

The wave and wind are random in nature which makes it challenging to accurately represent the load and load effects for offshore structures subjected to these loads. This randomness should be represented as accurately as possible to calculate reasonable hydro-aero-dynamic loading. Both the waves and the wind have long-term and short-term variability. Wave and wind are stochastic; and, their magnitude, directions and periods vary with time. The wave and wind are phenomena which transfer energy in water and air, respectively (considering marine/offshore technology terminology). The solar energy in earth makes air molecules to move, this motions appear as wind. Wind blowing over ocean makes boundary layer above sea surface and due to roughness of water surface the energy of wind transfers to wave particles making gradually waves which is so-called wind-generated-waves. To design, install and operate offshore energy structures in a safe, practical and cost-effective way, it is required to have realistic and reliable metocean (meteorological and oceanographic) data available from the offshore site for the conditions to which the installation may be exposed.

The required analysis time depends on the natural periods of the system. Wave-induced motions of common floating structures have been carried out considering 3-h short-term analysis. The 3-h simulation looks promising for such structures. However, the 10-min response analysis has been widely applied in wind engineering to cover all the physics governing a fixed wind turbine. When it comes to floating energy structures (e.g. FWTs), the correlation between the wave and wind should be accounted for. For each environmental condition, the joint distribution

© Springer International Publishing Switzerland 2014

M. Karimirad, *Offshore Energy Structures*, DOI 10.1007/978-3-319-12175-8_8

of the significant wave height, wave peak period, wave direction and mean water level (MWL; relevant for shallow water) combined with the mean wind speed, wind direction and turbulence should be considered (Karimirad 2011). In this chapter, wind and wave characteristics and applied theories in offshore engineering for these phenomena are discussed.

8.2 Regular Wave Theory

Ocean gravity waves with small amplitude compared to the wave length are usually described by linear potential wave theory (Faltinsen 1993) and (Newman 1977) The linear wave theory (usually called the Airy theory) can be used to represent the wave kinematics. In the regular wave theory, the wave is assumed to be sinusoidal with constant wave amplitude (ζ_a), wave length $\left(\lambda = \dfrac{2\pi}{k} \right)$ and wave period $\left(T = \dfrac{2\pi}{\omega} \right)$. The regular propagating wave (ζ) is defined as:

$$\zeta = \zeta_a \cos(kx - \omega t). \tag{8.1}$$

In which, ω is the wave frequency and k is wave number.

Airy theory is simple and widely applied in offshore technology. The theory is a first-order wave theory, also known as small-amplitude or linear wave theory. Airy theory has shown to give acceptable approximations of kinematics/dynamic properties of ocean waves; hence, it is often used to approximate wave behaviour in engineering applications.

In the Airy theory, the sea water is assumed to be incompressible and inviscid (nonviscous). If the fluid motion is assumed irrotational, then, a velocity potential exists (ϕ) and satisfies the Laplace equations $(\nabla^2 \phi = 0)$. Note that the assumptions made limit the Airy theory for large waves, i.e. in storm conditions. In the following, the governing equations and assumptions for Airy theory are described.

1. Laplace equations in 2-dimentional format:

$$\frac{\partial^2 \phi}{\partial x^2} + \frac{\partial^2 \phi}{\partial z^2} = 0 \tag{8.2}$$

$u_x = \dfrac{\partial \phi}{\partial x}$, $u_z = \dfrac{\partial \phi}{\partial z}$, in which, u_x and u_z are velocity of fluid in x and z directions, respectively.

By applying the kinematic boundary conditions and the dynamic free surface conditions, the velocity potential and the wave kinematics can be found. For small amplitudes, the boundary conditions can be linearized and the free surface wave elevation (ζ) is replaced with the undisturbed fluid surface where the velocity potential satisfies the boundary conditions (Water Waves 2011):

2. The linearized dynamic free surface boundary condition:

$$g\varsigma = -\frac{\partial \phi}{\partial t}\bigg|_{z=0} \qquad (8.3)$$

in which, g is gravitational acceleration (i.e. 9.81 m/sec^2).
The linearized kinematic free surface boundary condition:

$$\frac{\partial \phi}{\partial z}\bigg|_{z=0} = \frac{\partial \varsigma}{\partial t}. \qquad (8.4)$$

The bottom boundary condition specifies that the fluid velocity must be parallel to the seabed, as the water velocity normal to sea bottom should be zero (assuming that water will not penetrate the soil, impermeable soil).

$$\frac{\partial \phi}{\partial z}\bigg|_{z=-h} = 0 \qquad (8.5)$$

For the case of deep water, the velocity of flow dies at a depth below surface and seabed will not feel the ocean surface wave.

Based on the Airy theory, the velocity potential is derived as:

$$\phi = \frac{g\varsigma_a}{\omega} \frac{\cosh k(h+z)}{\cosh kh} \sin(kx - \omega t). \qquad (8.6)$$

In the derived equation for the velocity potential, the wave frequency, wave number and depth of water are dependent. The dispersion relation in terms of angular frequency and the wave number for a given depth should be satisfied which is expressed by:

$$\omega^2 = gk \tanh kh. \qquad (8.7)$$

The horizontal water particle kinematics, velocity and acceleration, are described by:

$$u_x = \omega \varsigma_a \frac{\cosh k(z+h)}{\sinh kh} \cos(kx - \omega t), \qquad (8.8)$$

$$a_x = \omega^2 \varsigma_a \frac{\cosh k(z+h)}{\sinh kh} \sin(kx - \omega t), \qquad (8.9)$$

where u_x and a_x are the water particle velocity and acceleration in the x-direction (wave propagation direction), respectively, ω is the wave frequency of the monochromatic wave, ς_a is the regular wave amplitude, k is the wave number, the z axis is upward, and h is the mean water depth. The dynamic pressure is presented below:

$$P_D = \rho g \varsigma_a \frac{\cosh k(z+h)}{\cosh kh} \cos(kx - \omega t). \qquad (8.10)$$

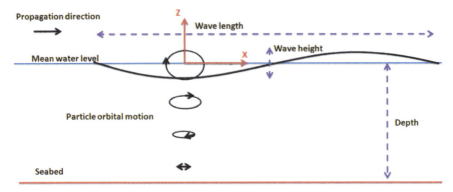

Fig. 8.1 Linear wave/regular wave theory (Airy theory) in finite depth

In deep water, the water particles move in circles in accordance with the harmonic wave. In deep water, the depth is greater than half of the wave length ($h \geq \lambda / 2$). Thus, the effect of the seabed cannot disturb the waves.

For shallow water, the effect of the seabed makes the circular motion into an elliptic motion (Karimirad 2011). At the seabed, for shallow water cases, the particles motion is completely horizontal due to large effect of sea bottom. Figure 8.1 shows a linear wave (Airy theory) in finite depth and its corresponding parameters.

One of the interesting wave parameter is the phase velocity which is the propagation velocity of the wave. It is also called wave speed or wave celerity and is determined by: $C = \lambda / T$. For linear waves, the phase velocity is expressed as:

$$C = \sqrt{\frac{g\lambda}{2\pi} \tanh\left(\frac{2\pi h}{\lambda}\right)}. \tag{8.11}$$

For deep water ($\omega^2 = gk$), the wave speed is $C = gT / 2\pi$ and for shallow water, the wave speed is $C = \sqrt{gh}$.

Example 1: For instance, a wave in deep water with period of 15 s has the wave length of 351 m. This means the water depth should be larger than 175.5 m and the wave speed is about 23.4 m/sec (84.3 km/hr).

The wave particle no longer forms closed orbital paths when the wave amplitude increases. This is due to the fact that, after the passage of each crest, particles are displaced from their previous positions. This phenomenon is known as the 'Stokes drift'. The Boussinesq equations that combine frequency dispersion and nonlinear effects are applicable for intermediate and shallow water. However, in very shallow water, the shallow water equations should be applied. A study of 'Nonlinear irregular wave forces on near-shore structures by a high-order Boussinesq method' for fixed-wind turbines in shallow water is presented in (Bingham and Madsen 2003). This kind of structure is exposed to irregular, highly nonlinear waves in intermediate to shallow depth water. Linear wave theory represents the first-order approximation in satisfying the free surface conditions. It is possible to improve the theory by introducing higher-order terms in a consistent manner by Stokes expansion method (Faltinsen 1993).

8.3 Modified Linear Wave Theory (Stretching Models)

The linear wave theory (Airy theory) is just valid up to the MWL surface $(z = 0)$ and does not describe the wave kinematics above that level. However, it is important to define wave kinematics above the water level surface up to the wave elevation to determine correctly the hydrodynamic loads. Different mathematical models, such as constant stretching and Wheeler stretching, have been suggested to describe the wave kinematics above the MWL surface (USFOS 2010).

1. In the constant stretching model, it is assumed that the wave kinematic is constant above the MWL. So, the wave kinematics at MWL is used for entire region above the MWL. This method is called 'extrapolated Airy theory'.

Figure 8.2 shows the dynamic pressure for extrapolated Airy theory, the dynamic pressure is assumed constant above MWL in wave crests while the 'true' value of dynamic pressure is used below MWL in wave troughs.

Figure 8.3 shows the static, dynamic and total pressure for Airy theory while uniform stretching approach is applied for over MWL. When increasing the depth,

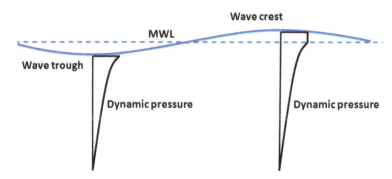

Fig. 8.2 Constant stretching, extrapolated Airy theory

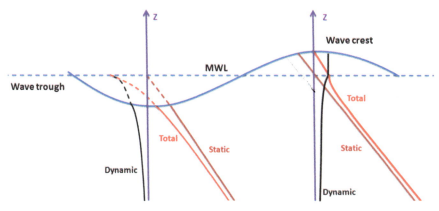

Fig. 8.3 Dynamic, static and total pressure for Airy theory (assuming uniform stretching for above MWL)

note that the static pressure increases linearly from MWL and the dynamic pressure decreases exponentially from MWL.

Based on the dynamic boundary condition, the static pressure should cancel the dynamic pressure when moving on the wave elevation. It is clear that the total pressure vanishes exactly at wave crest surface. However, there is a higher-order error (i.e. second-order error) associated to total pressure at the wave troughs; refer to (Faltinsen 1993).

2. In the Wheeler stretching model, the vertical coordinate (z) is substituted by the scaled coordinate (z').

$$z' = (z - \zeta)\frac{h}{h + \zeta},\qquad(8.12)$$

where $\zeta = \zeta_a \cos(kx - \omega t)$ is the wave elevation. It is possible to apply Wheeler stretching for irregular waves; the stretching in the stochastic context is explained later in this chapter.

When applying the uniform stretching, above the MWL, the dynamic pressure does not vary. Hence, if we have a submerged element, it will not be exposed to any dynamic vertical load (just static or buoyancy load presents). However, Wheeler stretching represents dynamic vertical force for submerged bodies regardless of vertical location (USFOS 2010; Fig. 8.4).

Wheeler stretching and uniform constant stretching has similar wave kinematics at MWL. This means for offshore structures in which inertia loads are governing, the two theories agree as the maximum acceleration terms occur when the wave elevation is around MWL. Example of such structure is a spar-type wind turbine while for drag dominated structures such as jacket wind turbines, these theories have different load and load effects. This is due to the fact that the maximum/minimum drag force occurs at wave crest and trough (maximum/minimum velocities) where wave kinematic from these stretching approaches is not consistent.

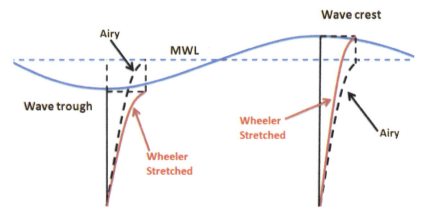

Fig. 8.4 Wheeler stretching

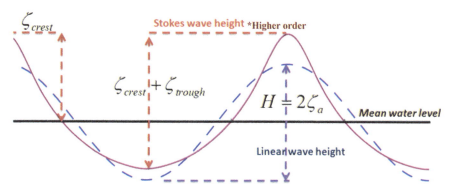

Fig. 8.5 Stokes wave, linear and higher order, the maximum crest/height ratio for a Stokes wave is 0.635

8.4 Stokes Wave Theory

The Stokes wave theory is based on expansion of the surface elevation in powers of the linear wave height. Figure 8.5 illustrates Stokes waves, a compression between linear (first order) and higher-order wave elevation.

- First order:

The first-order Stokes wave is identical to Airy wave (linear wave) which has been discussed in the previous sections.

- Second order:

The surface elevation profile for a regular second-order Stokes wave is defined by:

$$\zeta = \frac{H}{2}\cos(kx - \omega t) + \frac{\pi H^2}{8\lambda}\frac{\cosh kh}{\sinh^3 kh}(2 + \cosh 2kh)\cos 2(kx - \omega t) \quad (8.13)$$

$$H = \zeta_{max} - \zeta_{min}$$
$$H = 2\zeta_a \quad \text{for Airy wave}$$

Wave elevation ζ is written in two-dimensional as a function of time (t) and location (x); hence appearing as $\zeta(x,t)$. The three-dimensional format of wave elevation $\zeta(x,y,t)$ is a general presentation, in which $\Theta = (kx - \omega t)$ is replaced by $\Theta = k(x\cos\chi + y\cos\chi) - \omega t$; χ is the direction of propagation, measured from the positive x-axis. This means the first-order wave in three-dimensional format is $\zeta = \zeta_a \cos(k(x\cos\chi + y\cos\chi) - \omega t)$.

In deep water, the Stokes second-order wave is given by

$$\zeta = \frac{H}{2}\cos(kx - \omega t) + \frac{\pi H^2}{4\lambda}\cos 2(kx - \omega t). \quad (8.14)$$

The crests are steeper and troughs are wider in second- and higher-order Stokes waves compared to linear waves. For example, the crest height is increased by factor of $1+(\pi H/2\lambda)$ compared to linear Airy wave. The crest and trough elevation of a second-order Stokes wave are:

$$\zeta_{crest} = \zeta|_{\Theta=0} = \frac{H}{2}+\frac{\pi H^2}{4\lambda} = \frac{H}{2}\left(1+\frac{\pi H}{2\lambda}\right)$$

$$\zeta_{trough} = \zeta|_{\Theta=\pi} = -\frac{H}{2}+\frac{\pi H^2}{4\lambda} = \frac{H}{2}\left(\frac{\pi H}{2\lambda}-1\right). \tag{8.15}$$

For linear wave: $\dfrac{\zeta_{crest}}{H} = \dfrac{1}{2}+\dfrac{\pi H}{4\lambda}\bigg|_{\frac{H}{\lambda}\to 0} = 0.5.$

For second-order Stokes wave: $\dfrac{\zeta_{crest}}{H} = 0.5+\dfrac{\pi H}{4\lambda}.$

Example 2: Consider the same wave mentioned in Example 1. If the wave height for linear wave is considered to be 10 m, the second-order Stokes wave theory results in $\dfrac{\zeta_{crest}}{H} = 0.5+\dfrac{\pi\times10}{4\times351} = 0.5+0.02 \approx 0.52.$ In shallow water, higher-order Stokes waves (e.g. fifth order) results in larger crest/height ratio. The maximum crest to wave height ratio for a Stokes wave is 0.635.

- 3rd order:

In deep water, the Stokes third-order wave is given by:

$$\zeta = \frac{H}{2}\cos(kx-\omega t)+\frac{\pi H^2}{4\lambda}\cos 2(kx-\omega t)+\frac{3H^3\pi^2}{16\lambda^2}\cos 3(kx-\omega t) \tag{8.16}$$

$$\frac{H}{2}\cos(kx-\omega t): \quad First-order$$

$$\frac{H}{2}\cos(kx-\omega t)+\frac{\pi H^2}{4\lambda}\cos 2(kx-\omega t): \quad Second-order$$

The linear dispersion relation holds for second-order Stokes waves. The wave height is modified in second-order Stokes theory; however, the phase velocity and the wave length remain independent of wave height. In third order, the phase velocity depends on wave height:

$$c = \sqrt{\frac{g}{k}\tanh(kh).\left[1+\left(\frac{kH}{2}\right)^2.\left(\frac{9-8\cosh^2(kh)+8\cosh^4(kh)}{8\sinh^4(kh)}\right)\right]}. \tag{8.17}$$

For deep water, it simplifies to:

$$c = \sqrt{\frac{g}{k}\left(1+\left(\frac{kH}{2}\right)^2\right)} \qquad (8.18)$$

• Fifth order:

The fifth-order Stokes wave theory gets complicated due to added extra terms in wave elevation extension and added mathematical calculations. Iterative solution is needed to define the wave elevation in this case. The wave potential is given by a series expansion with five terms, for instance refer to (Brorsen 2007):

$$\Phi = \sum_{i=1}^{5} \phi_i \cosh(k(z+h))\cos(\omega t - kx). \qquad (8.19)$$

In which:

$$
\begin{aligned}
\phi_1 &= \lambda A_{11} + \lambda^3 A_{13} + \lambda^5 A_{15} \\
\phi_2 &= \lambda^2 A_{22} + \lambda^4 A_{24} \\
\phi_3 &= \lambda^3 A_{33} + \lambda^5 A_{35} \qquad A_{ij}\ _{\substack{for\,i=1:5 \\ for\,j=1:5}} = function(kh) \\
\phi_4 &= \lambda^4 A_{44} \\
\phi_5 &= \lambda^5 A_{55}
\end{aligned} \qquad (8.20)
$$

$$\omega^2 = gk\tanh(kh).(1+\lambda^2 C_1 + \lambda^4 C_2) \quad C_{ij}\ _{\substack{i=1:5 \\ j=1:5}} = function(kh) \qquad (8.21)$$

8.5 Cnoidal and Solitary Wave Theories

Sharp crests and very flat troughs are characteristics for cnoidal waves. The cnoidal wave is a periodic wave with sharp crests separated by wide troughs. The cnoidal wave theory for the steady water wave problem follows from a shallow water approximation. Cnoidal wave theory is used when shallow water parameter ($\mu = h/\lambda_0$) < 0.125 and Ursell number ($U_R = H\lambda^2/h^3$) > 30 (DNV 2007) For more information refer to (Fenton 1990). A cnoidal wave has crest/height ratio between 0.635 and 1 (note: the maximum crest/height ratio for a Stokes wave is 0.635).

Solitary wave is a special case in which $\zeta_{crest} = H$. For high Ursell numbers, the wave length of the cnoidal wave goes to infinity and the wave is a solitary wave. A solitary wave is a propagating shallow water wave. The surface elevation of solitary wave is above the MWL; refer to (Fenton 1972, 1979).

8.6   Stream Function Wave Theory

The basic form of the Stokes approach is a Fourier series which is capable of accurately approximating any periodic quantity if the coefficients in that series can be found. A procedure can be based on perturbation expansions for the coefficients in the series as it is done in Stokes theory, or based on calculation of the coefficients numerically by solving the full nonlinear equations. The second approach is implemented behind the methods presented in (Chappelear 1961), (Dean 1965), and (Rienecker and Fenton 1981). Chappelear implemented the velocity potential for the field variable and introduced a Fourier series for the surface elevation.

Stream function wave theory was developed by Dean; refer to (Dean 1965). Dean applied the stream function for the field variable and point values of the surface elevations to obtain a rather simple set of equations, so-called 'stream function wave theory'. Stream function wave theory covers a broader range of applicability than the Stokes wave theory. Dean applied a series solution in sine and cosine terms to the fully nonlinear water wave problem involving the Laplace equation with two nonlinear free surface boundary conditions, constant pressure, and a wave height constraint.

The stream function order is related to the wave nonlinearities. To accurately present the waves that are close to breaking wave height limits, higher-order stream function with more terms are required. Third to fifth order may be enough to model waves in deep water while high order, i.e. 30th order is needed for waves in very shallow water. A convergence and sensitivity studies are required to measure which order is needed for a particular offshore site (if the changes are negligible, the order used is enough). If the wave height to depth ratio is less than 0.5, the Stokes wave theory (fifth order) and stream function wave theory have a good agreement (USFOS 2010). Stream function wave theory should be used in very shallow water. For more information regarding the nonlinear wave and advanced theories refer to literature such as (Fenton 1990), (Brorsen 2007).

8.7   Validity Range of Wave Theories

The wave theory discussed here, Airy theory in its original format, is just applicable for nonbreaking waves. The wave breaks (DNV 2007):

1. in shallow water when $H / h \geq 0.78$
2. and in deep water when $H / \lambda \geq 0.14$

where, H is wave height, λ is wave length and h is mean water depth.

For harsh conditions and extreme waves, when the height increases, linear wave theory cannot capture the nonlinear features of the wave kinematics. The existent nonlinear methods are mainly applicable for deterministic waves (regular waves). These nonlinear methods are not suitable for stochastic wave fields. The good news is that the probability of breaking waves is relatively small; and, most of the waves break close to the coast and not at the offshore site. However, the breaking waves, their probability of occurrences and corresponding load-effects in offshore structures would need to be considered for detailed design.

To determine the validity range of a specific wave theory for a given offshore site, the regular wave parameters such as wave height and period as well as water depth are studied. Three nondimensional parameters are defined for this purpose (DNV 2007):

$S = H / \lambda_0$: Wave steepness parameter

$\mu = h / \lambda_0$: Shallow water parameter

$U_R = H \lambda^2 \big/ h^3$: Ursell number

In which, λ_0 is the deep water wave number corresponding to the deep water dispersion relation ($\omega^2 = gk$). For more information refer to DNV-RP-C205, Environmental Conditions and Environmental Loads.

Ranges of validity for various wave theories based on these nondimensional parameters are determined. An example of such a plot defining the ranges can be found in 'Hydrodynamics of Offshore Structures', WIT Press by Chakrabarti S.K. (1987). In the coming sections, more wave theories are explained.

For regular steep waves, the fifth-order Stokes wave theory applies. Stokes wave theory is not applicable for very shallow water. Hence, cnoidal wave theory or stream function wave theory should be used. If Ursell number is around 30, both Stokes wave theory and cnoidal wave theory have inaccuracies and the stream function method is recommended by standards, i.e. refer to 'Environmental Conditions and Environmental Loads' by (DNV 2007).

8.8 Offshore Waves Versus Nearshore Waves

Here, some phenomena that happen for water wave when wave enters shallow water compared to deep water are mentioned. The wave speed decreases as waves approach the shoreline. Also, the waves pinch together at the front and the wave steepens, the height of the wave increases, and eventually the wave breaks (in shallow water when $H / h \geq 0.78$).

Total energy of the wave is a sum of kinetic and potential energies (the energy of the water wave is transmitted as kinetic and potential energies). The kinetic energy is related to the motion of the water (movement of the water particles, i.e. circular movement in the linear wave theory). The potential energy is related to the gravitational potential energy, i.e. the amount of water exceeding the MWL (and the crest height).

The water particles at the deeper part of the water store the kinetic energy when the waves are in moderate and deep water depth. But, when the wave enters shallow water, the circles on the bottom of the wave are compressed. The motion of the water stopped due to the effect of the seabed and the kinetic energy of this part is decreased. Even though, some of the wave energy is lost due to friction to the sea bottom, a lot of it is conserved as gravitational potential energy. Hence, the height of the waves increases as they enter shallow water. In another word, the kinetic energy of the water near the bottom is converted to potential energy represented as wave height.

In shallow water, the wave velocity is determined by the depth. The water particles at the top of the wave are less affected by the seabed. Hence, the top water moves faster than the bottom water and the wave front steepens in shallow water due to seabed effects. Ultimately, the top water moves so far relative to the bottom water that the wave breaks.

8.9 Irregular Wave Theory

Both swell and wind-generated waves present in offshore sites. Swell is generally regular wave caused by large meteorological phenomena. Swell continues after the source has disappeared and maintains its direction as long as it is in deep water. Usually, swell travels for a considerable distance. Ocean wind-generated waves are stochastic in nature. The ocean waves are mostly presented as combination of regular waves, forming irregular waves.

Wind-generated waves are sustained by receiving energy from the wind. This means they disappear if the wind stops blowing over ocean water. However, swell waves are free waves and after forming in their origin, they travel to new sites where they do not receive wind energy.

In ocean engineering, it is usually assumed that the sea surface forms a stochastic wave field that is stationary in a short term period, i.e. in 3-h intervals. The stationary assumption of the wave is offshore-site-dependent. This approach is well-documented and -tested in the past years. The stationary assumption of the wave over a defined period works very well and gives good agreement with full-scale measurements for most practical offshore engineering cases.

In most marine technology cases, Gaussian assumption of the wave field gives a reasonably good approximation of reality. Such assumption for stochastic model of waves leads to a normally distributed water surface elevation. If the wave elevation is assumed to be Gaussian and narrow-banded, the wave crests follow the Rayleigh distribution. For most wave and wave-effects, the Gaussian assumption presents a reliable and accurate model. But, some phenomena such as slamming do not have a Gaussian distribution.

The stationary sea represented by the wave elevation at a point in space is modelled by a wave spectrum applying the Gaussian assumption. Different models have been proposed for wave spectra depending to condition of ocean waves. For example, (1) the Pierson-Moskowitz (PM) and Joint North Sea Wave Project (JON-SWAP) spectra are two known mathematical models for wind-generated waves. The PM spectrum is proper for a fully developed wind sea while for a growing wind sea, (2) the JONSWAP spectrum is recommended. (3) Torsethaugen spectrum (two-peaked wave spectrum) is introduced to define a sea comprising wind-generated waves and swells, refer to (Knut and Sverre 2004). There are other spectra such as: (4) International Ship and Offshore Structures Congress (ISSC) spectrum also known as Bretschneider or modified Pierson-Moskowitz and (5) Ochi-Hubble spectrum (for sea states that include both swell and wind-generated waves). Ochi and Hubble suggested a six-parameter spectrum with a superposition of two modified Pierson-Moskowitz spectra; refer to (Ochi and Hubble 1976).

• PM Spectrum

$$S_{PM}(f) = \frac{5}{16} \frac{H_S^2}{T_P^4 f^5} \exp\left(-\frac{5}{4}(fT_P)^{-4}\right), \tag{8.22}$$

where H_S is the significant wave height, T_P is the wave peak period, and f is the frequency in hertz $(f = \omega / 2\pi)$.

• JONSWAP Spectrum

$$S_{JS}(f) = F_n S_{PM}(f) \gamma_{JS} \exp\left(\frac{-(f - f_P)^2}{2\sigma_{JS}^2 f_P^2}\right) \tag{8.23}$$

$$\sigma_{JS} = \begin{cases} \sigma_a & \text{for } f \le f_P \text{ (typically :0.07)} \\ \sigma_b & \text{for } f > f_P \text{ (typically :0.09)} \end{cases}$$

$$F_n = 5(0.065\gamma_{JS}^{0.803} + 0.135)^{-1} \quad \text{for} 1 \le \gamma_{JS} \le 10$$

where γ_{JS} is the shape parameter, which, for seas that are not fully developed is around 3.3. For fully developed seas, γ_{JS} is taken to be 1. Therefore, the JONSWAP and PM spectra are the same for $\gamma_{JS} = 1$. $f_P = \frac{1}{T_P}$ is the wave peak frequency in hertz.

When the wave spectrum for operational and harsh environmental conditions are compared, it is found that the extreme sea state has much larger peaks, and it also covers a wider range of frequencies. The peak frequency of a harsh sea state is shifted to lower frequencies as well. This shift means that the probability of resonant motion occurrence for floating structures with slowly varying motions is higher in extreme environmental conditions (Karimirad 2011).

In Fig. 8.6 a wave spectrum and a realization of an irregular wave is shown. The wave elevation at a given point is a superposition of several regular waves; each of them is in the format of $\zeta = \zeta_a \cos(kx - \omega t)$. When, they are summed up phase angle (ϕ) presents to make them individual, remembering that the wave field is a stochastic phenomena. At a given frequency, regular wave amplitude has a relation to wave spectrum area. The area under the wave spectrum represents the energy of the wave and hence, it is easy to relate it to wave amplitude.

In other words, to start a wave kinematic calculation for a suitable wave spectrum representing the offshore site, the computations begin by converting the spectrum back into individual sinusoids. This is done through different strategies based on

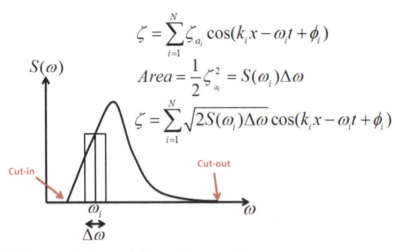

$$\zeta = \sum_{i=1}^{N} \zeta_{a_i} \cos(k_i x - \omega_i t + \phi_i)$$

$$Area = \frac{1}{2}\zeta_{a_i}^2 = S(\omega_i)\Delta\omega$$

$$\zeta = \sum_{i=1}^{N} \sqrt{2S(\omega_i)\Delta\omega} \cos(k_i x - \omega_i t + \phi_i)$$

Fig. 8.6 An example of ocean wind-generated wave spectrum

fast Fourier transforms (FFT) and inverse fast Fourier transform (IFFT) approaches. If the position (due to large motions, e.g. for a floating wind turbine) is updated, the FFT algorithm is not directly applicable, and the discrete Fourier transform (DFT) is usually applied.

The total number of regular waves (N) added to make an irregular wave should be enough to avoid repetition of the irregular wave, i.e. roughly 1000 regular wave frequencies are needed for each 1-h analysis.

Each sinusoid (regular) wave has a frequency and amplitude. The amplitude as mentioned above is derived from the energy density defined by the spectrum of ocean waves. Usually, the wave filed and related kinematics is pregenerated before hydrodynamic load calculation and dynamic analysis due to heavy computations demanded.

The wave spectrum is used to represent the energy distribution for each frequency. For a specific offshore site, the significant wave height and wave peak period are identified. Depending to the offshore site, a proper wave spectrum is selected. Usually, for developing wind-generated waves, the JONSWAP spectrum is suitable for the North Sea. Based on the significant wave height and the peak wave period, the distribution of energy for each frequency is defined for the chosen characteristics (Karimirad 2011).

It is possible to apply stretching, i.e. Wheeler for irregular wave field. In the stochastic context, for each regular wave involved, the stretching is separately applied. Then, the wave kinematics up to the instantaneous water level is obtained for each regular wave participating in the superposition. After that, the wave kinematic of regular waves are added together to obtain the irregular wave kinematic (i.e. wave elevation, velocity and acceleration) for a particular time. In Chap. 11, the stochastic methods and theories are explained in more details.

8.10 Wind Theory

In the following sections, the practical wind theories applied for offshore energy structures are described. The important wind characteristics needed for a proper wind field in offshore wind technology are defined. Aerodynamic and hydrodynamic loads are described in the next chapter.

8.11 Spatial and Temporal Variations of Wind

Wind has a stochastic nature (same as ocean waves) and is characterized by its speed and direction. Turbulent wind theory is explained herein. Wind varies over space and time, spatial and temporal variation of wind should be accounted for when calculating the wind loads and its corresponding load effects. These variations are affecting the power performance and structural integrity of wind devices. To investigate the site energy resource which is the first concern for a specific location, turbulent wind field should be evaluated. Spera (1998) has mentioned the spatial and temporal variations of the wind; it is possible to summarize them as following.
Spatial variations:

- Trade winds emerging from subtropical, anticyclonic cells in both hemispheres.
- Monsoons which are seasonal winds generated by the difference in temperature between land and sea.
- Westerlies and subpolar flows.
- Synoptic-scale motions which are associated with periodic systems such as travelling waves.
- Mesoscale wind systems which are caused by differential heating of topological features and called breezes.
- Temporal variations of wind appear as the following forms:
- Long term variability which are annual variations of wind in a special site
- Seasonal and monthly variability
- Diurnal and semidiurnal variation
- Turbulence (range from seconds to minutes)

The temporal variations are usually represented by the energy spectrum of the wind, i.e. refer to the Van der Hoven wind speed spectrum (Hoven 1957). Wind spectrum has two main peaks: one peak corresponds to 4-day period and the other one corresponds to 30-s period. The yearly wind speed variations, pressure systems and diurnal changes are influencing the left side of the wind speed spectrum corresponding to 4-day peak. However, the turbulence shows itself in the right side of the spectrum, influencing the 30-s peak, see Fig. 8.7.

As it is shown in Fig. 8.7, the wind energy is concentrated around two separated time-periods (diurnal and 1-min periods). This allows the splitting of the wind speed into two terms:

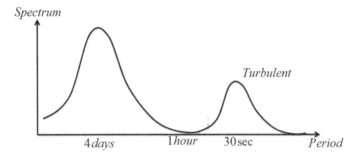

Fig. 8.7 Schematic of a wind spectrum accounting for long-term (diurnal) and short-term (turbulence) temporal variations

- Quasi-steady wind speed (usually called the mean wind speed), V_{mean}
- Dynamic part (the turbulent wind), V_{turb}

$$V(t) = V_{mean} + V_{turb}(t). \tag{8.24}$$

Time-varying wind speed $(V(t))$ consists of steady value (V_{mean}) plus the fluctuations $(V_{turb}(t))$ about the steady value. For probabilistic response analysis of off-shore energy structures, it is practical and accurate enough to assume steady part of the wind as quasi-static; and, neglect its time dependency. This simplifies the wind modelling and numerical simulation of wind field to a large extent.

Mean wind speed (V_{mean}) is usually referred to an averaging period, i.e. 10-min averaged, V_{10min}. As it is discussed above and is clear in Fig. 8.7, there is an energy gap in the wind spectrum, roughly between 20-min to couple of hours. This means the quasi-static assumption of mean wind speed can remain for the time averaging more than 10-min (even up to the spectrum gap, i.e. couple of hours). However, if the metocean data is based on a time-averaging which is different with the needed data-format for the analysis input, then, a scaling factor should be accounted for this conversion. The ratio between 10-min mean wind speed and 1-h mean wind speed at the Brent Statfjord site is around 1.085, refer to (Karimirad and Moan 2011).

8.12 Wind Distribution (Weibull Long-Term Probability)

Site measurements of wind velocity are fitted to Weibull distribution in order to derive a long-term probability distribution of the mean wind speed. This long-term probability distribution is used to assess the energy resource of the offshore site as well as metocean input data for structural integrity assessment of the energy devices. The Weibull probability density function (f_W) of the wind speed is defined by the following formula:

$$f_W(V) = \frac{k_w}{c_w} \left(\frac{V}{c_w} \right)^{k_w - 1} \exp\left(-\left(\frac{V}{c_w} \right)^{k_w} \right). \tag{8.25}$$

where, V is the wind speed, k_w is the shape-parameter describing the variability about the mean, and c_w is a scale-parameter related to the annual mean wind speed. The moderate winds are more frequent than the high-speed winds and this is clear in Weibull distribution. The shape and scale parameters of Weibull probability distribution are site dependent and vary for different areas, i.e. onshore, coastal and offshore. If $k_w = 2$, a special case appears in which mean wind velocity has a Rayleigh distribution.

$$f_W(V) = \frac{2V}{c_w^2} \exp\left(\frac{-V^2}{c_w^2}\right) \qquad (8.26)$$

To assess the annual power resources of a site, Lysen defined an empirical formula for the annual mean wind speed: $V_{Annual} = c_w(0.568 + 0.433/k_w)^{1/k_w}$, refer to (Lysen 1983).

8.13 Wind Shear

Due to roughness of surface over which the wind blows, the wind speed varies respect to height and the mean wind speed is a function of height. Mathematical models are represented to account for this variation which is so-called wind shear. Different shear models such as the Prandtl logarithmic and power laws are usually applied in wind turbine technology. In these mathematical models (log law and the power law), a parameter called the roughness length or exponent includes the effect of the surface type over which the wind blows.

The wind speed is usually referred to a reference point, i.e. 10 m above MWL surface. The shear models are used to describe the wind velocity at any height. The logarithmic law is described by the following formula:

$$V(z) = V(h) \frac{\ln\dfrac{z}{z_0}}{\ln\dfrac{h}{z_0}} \qquad (8.27)$$

where $V(z)$ is the wind speed at the height of z, h is the reference height (10 m in most cases), and z_0 is the roughness parameter, which depends on the wind speed, distance to the land, water depth, and wave field of offshore sites. It varies from 0.0001 for calm sea to 0.003 in coastal areas with onshore wind (Karimirad and Moan 2011) Roughness parameter can be expressed by the following equation in which, κ is von-Karman's constant and A_C is Charnock's constant varies from 0.011 for open sea to 0.034 for near coastal areas.

$$z_0 = \frac{A_C}{g}\left(\frac{\kappa V_{10min}}{\ln\dfrac{z}{z_0}}\right)^2 \qquad (8.28)$$

Power law wind shear model is an empirically developed relationship given as

$$V(z) = V(h)\left[\frac{z}{h}\right]^{\alpha} \tag{8.29}$$

The power law exponent is α. For the fairly flat terrain, $\alpha = 1/7$ and for the ocean surface, $\alpha = 1/10$ is proposed. In most cases, the difference between obtained wind speed based on the power and logarithmic shear laws is small, i.e. 1 %.

8.14 Turbulence and Wind Spectrum

Wind is the moving air particles with a dominant velocity and direction. Air particles carry kinetic energy. The wind kinetic energy converts to thermal energy due to creation and destruction of progressive smaller eddies and gusts. This dissipation of wind energy causes turbulence in the wind field. Wind is a turbulent phenomenon in nature. The good thing is that over time periods of an hour and more, the wind has a relatively constant mean. Otherwise, in shorter periods, i.e. minutes, it is quite variable.

Turbulence is the dynamic part of the wind speed including all wind speed fluctuations with periods below the spectral gap. As it is explained above, the spectral gap occurs around 1-h separating the slowly-varying and turbulent ranges. Hence, all spectral components in the range from seconds to minutes are accounted in the turbulence. Turbulent wind is three-dimensional, consisting longitudinal, lateral, and vertical components.

To quantify the turbulent, the ratio of standard deviation of the wind speed (σ) over the mean wind speed (μ) for a specified time period (which is normally less than 1 h) is used. This ratio ($I = \sigma/\mu$) is called 'turbulence intensity'. The time period for defining the turbulence intensity is usually 10 min for onshore wind technology. The turbulence intensity is a function of wind speed and in standards depending to the class, it varies. Class 'C' is dedicated to offshore conditions.

In general, the turbulence intensity decreases with height. The turbulence intensity is higher when there are more obstacles in the terrain; therefore, the offshore turbulence is less than onshore turbulence intensities. As an example: for an offshore application, the turbulence intensities can be 0.10 and 0.15 for survival and operational conditions, correspondingly.

The captured annual power is not significantly affected by turbulence while the turbulence has a major impact on the structural integrity and power performance of the energy device. Aero-elastic dynamic responses of the slender parts of the structure such as blades are highly affected by the turbulent wind.

Power spectrum is used to describe the wind turbulence in a given point in space. The Kaimal and von Karman spectra are extensively applied in wind application. The turbulent wind spectrum is a function of the frequency, turbulence intensity, topography of the environment and mean wind speed. The Kaimal spectrum is defined by:

$$S(f) = \frac{I_t^2 V_{10\min} l}{\left(1 + 1.5 \dfrac{fl}{V_{10\min}}\right)^{5/3}} \tag{8.30}$$

where $I_t = {}^{\sigma}/_{V_{10\min}}$ is the turbulent intensity, f is the frequency in hertz, $V_{10\min}$ is the 10-min averaged wind speed, and l is a length scale. $l = 20h_{agl}$ for $h_{agl} < 30m$ or $l = 600m$ otherwise, where h_{agl} is the height above ground level (Hansen 2008).

The wind is more turbulent in harsh conditions compared to operational conditions and the spectrum under harsh conditions has much more energy compared to those under operational conditions. Especially, in the low frequency region, the wind spectrum corresponding to storm cases has a significant value.

At onshore sites, the obstacles in the terrain influence the boundary layer and make the wind more turbulent while at the offshore site, the wind is steadier, and the turbulence is decreased. Hence, the turbulence intensity of the offshore wind is less compared to onshore sites. In Fig. 8.8, a schematic of a wind spectrum (turbulence) is compared to a wave spectrum. The wind spectrum, especially for operational wind cases, covers the low frequency region, i.e. 0.0–0.5 rad/sec. In general, the main energy of the turbulent wind is concentrated below 0.3 rad/sec while waves are normally covering a higher range, i.e. 0.3–1.0 rad/sec. Floating wind turbines and hybrid marine platforms may have low natural frequencies which can be excited by wind loads. So, the turbulent wind loading is relatively less influencing the global responses of land-based wind turbines compared to offshore wind turbines in this respect.

Another turbulence model which is implemented in advanced aero-elastic codes such as Horizontal Axis Wind turbine (HAWC2) (Larsen and Hansen 2008) is Mann uniform shear turbulence model for generating three-dimensional wind field (Mann 1994). In Mann model, the turbulent velocity fluctuations are assumed to be a stationary, random vector field. The components of turbulent velocities assumed to have zero-mean Gaussian statistics.

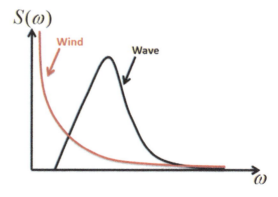

Fig. 8.8 Schematic of a wind spectrum versus wave spectrum

The description of Mann model is different with the other models in that a three-dimensional velocity spectral tensor is defined. The model assumes that the isotropic von Karman energy spectrum is rapidly distorted by a uniform, mean velocity shear. The Mann turbulence model is defined usually in wave number as (Karimirad and Moan 2011):

$$S(k_1) = 0.4754\sigma_{iso}^2 l^{-2/3} k_1^{-2/3}$$ (8.31)

k_1 is the nondimensional spatial wave numbers for the horizontal direction. l is the length scale: $(0.7\Lambda - 0.8\Lambda)$, in which Λ is the longitudinal turbulence scale parameter and defined based on hub-height z, as following:

$$\Lambda = \begin{cases} 0.7z & z \leq 60m \\ 42m & z \geq 60m \end{cases}$$ (8.32)

$\sigma_{iso} = 0.55\sigma_1$, σ_{iso} is the unsheared, isotropic variance and σ_1 is the turbulence standard deviation.

8.15   Joint Wave and Wind Conditions

For some of offshore energy structures, especially for those that have wind turbine included, the joint probability of the wave and wind should be considered when carrying out the stochastic analyses such as fatigue and extreme load effect evaluations. The wave and wind show long-term and short-term variability. The long-term variability of the wind can be defined by the mean wind speed and direction. The short-term variability of the wind is usually defined by the turbulence, as we have already discussed.

It is mentioned earlier that waves are usually wind-generated at offshore sites. In an offshore site, the ocean waves can be wind-generated and swell. This means there is a correlation between wave characteristics (the waves are usually defined by the peak period and significant wave height) and wind parameters. We learnt that the turbulence is a function of mean wind velocity. Moreover, the significant wave height is related to mean wind speed. In ocean engineering, the relation between wave period and significant wave height through scatter diagrams are well defined.

Proper site assessments reporting metocean data are required for setting the joint distribution of waves and wind prior to the analysis. Normally, the metocean data includes wave and wind characteristics such as the mean wind speed, turbulence, direction of the waves and wind, significant wave height, and wave peak period.

The development of the joint distribution requires measurement of simultaneous wave and wind time histories at the offshore sites for several years. However, limited site assessments having the correlated wave and wind time series are available. Also, current metocean data are usually got for oil and gas projects and hence, these data are missing the correlation between the turbulence and wave/wind characteristics. Hybrid marine platform is a new offshore technology. In future, accurate metrological and oceanological studies for determining proper environmental characteristics including the joint distribution of wave and wind should be set.

When considering the metocean data for offshore wind turbines and hybrid marine platforms, the following points should be considered.

1. Directionality of the wave/wind and possible misalignment
2. Relation between mean wind speed and turbulent intensity
3. Relation between mean wind speed and significant wave height

In general, the significant wave height increases with the increase of the wind speed. For higher wind speeds, the Weibull distribution is negatively skewed. For each wind speed, a range of significant wave heights is possible. Smaller wind speeds have a narrower range of significant wave heights. The IEC 61400-3 standard recommends the use of the median significant wave height at each wind speed for dynamic response analysis of offshore wind turbines (Karimirad 2011).

4. The correlation between mean wind speed, significant wave height, and wave peak period

Fitting the analytical functions to the site assessments by considering a mathematical distribution for the mean wind speed and significant wave height is an option; refer to (Johannessen et al. 2001).

References

Bingham, H. B., & Madsen, P. A. (2003). *Nonlinear irregular wave forces on near-shore structures by a high-order Boussinesq method*. 18th IWWFB. Le Croisic (France): 18th IWWFB.

Brorsen, M. (2007). *Non-linear waves*. Denmark: Aalborg University. ISSN:1901-7286

Chakrabarti, S. K. (1987). *Hydrodynamics of offshore structures*. WIT press.

Chappelear, J. E. (1961). Direct numerical calculation of wave properties. *Journal of Geophysical Research, 66*, 501–508.

Dean, R. G. (1965). Stream function representation of nonlinear ocean waves. *Journal of Geophysical Research, 70*, 4561–4572.

DNV. (2007). *Environmental conditions and environmental loads*. Norway: RECOMMENDED PRACTICE, DNV-RP-C205.

Faltinsen, O. (1993). *Sea loads on ships and offshore structures*. UK: Cambridge University Press.

Fenton, J. D. (1972). A ninth order solution for the solitary wave. *Journal of Fluid Mechanics, 53*, 257–271.

Fenton, J. D. (1979). A high order cnoidal wave theory. *Journal of Fluid Mechanics, 94*, 129–161.

Fenton, J. D. (1990). Nonlinear wave theories. *The sea volume 9: Ocean engineering science*. New Zealand.

Hansen, M. O. (2008). *Aerodynamics of wind turbines* (2nd ed.). UK: Earthscan.

Hoven, I. v. (1957). Power spectrum of horizontal wind speed in the frequency range of 0.0007–900 cycles per hour. *Journal of Metrology, 14*(2), 160–164.

Johannessen, K., Meling, T. S., & Haver, S. (2001). *Joint distribution for wind and waves in the northern North Sea*, ISOPE.

Karimirad, M. (2011). *Stochastic dynamic response analysis of spar-type wind turbines with catenary or taut mooring systems*. PHD thesis, NTNU, Norway.

Karimirad, M., & Moan, T. (2011). Extreme dynamic structural response analysis of catenary moored spar wind turbine in harsh environmental conditions. *Journal of Offshore Mechanics and Arctic Engineering, 133*(4), 041103.

Knut, T., & Sverre, H. (2004). *Simplified double peak spectral model for ocean waves.* ISOPE, (Paper No. 2004-JSC-193), France.

Larsen, T. J., & Hansen, A. M. (2008). *HAWC2 user manual.* Denmark: DTU.

Lysen, E. H. (1983). *Introduction to wind energy.* SWD Publications, SWD 82-1: The Netherlands.

Mann, J. (1994). The spatial structure of neutral atmospheric surface-layer turbulence. *Journal of Fluid Mechanics, 273,* 141–168.

Newman, J. N. (1977). *Marine hydrodynamics.* USA: MIT Press.

Ochi, M. K., & Hubble, E. N. (1976). *On six-parameters wave Spectra* (pp. 301–328). Proceedings of 15th Coastal Engineering, Honolulu, Hawaii.

Rienecker, M. M., & Fenton, J. D. (1981). A Fourier approximation method for steady water waves. *Journal of Fluid Mechanics, 104,* 119–137.

Spera, D. A. (1998). *Wind turbine technology fundamental concepts of wind turbine.* New York: ASME press.

USFOS. (2010). USFOS, hydrodynamic, theory description. Norway. http://www.usfos.no/manuals/usfos/theory/documents/Usfos_Hydrodynamics.pdf. Accessed Aug 2013.

Water Waves. (2011). Department of Mechanical Engineering, MIT Marine Hydrodynamics course, Lecture 14, Chapter 6 - Water Waves. http://web.mit.edu/2.20/www/lectures/lec14/lecture14.pdf. Accessed Nov 2014.

Chapter 9
Aerodynamic and Hydrodynamic Loads

9.1 Introduction

Offshore energy structures are subjected to oceanic environmental loads. Aerodynamics and hydrodynamics are the governing loads for the majority of the structures and structural components. Generally, the most important hydrodynamic and aerodynamic loads are presented by wave and wind loading. However, in some cases and for special design/concepts, the ocean current and hydrostatic loads may significantly affect the scantlings (Fig. 9.1).

This chapter studies the main hydrodynamic and aerodynamic loads keeping an eye on the offshore energy structure applications. The aerodynamics are discussed with emphasige on wind turbine and its corresponding load and load effects on the main parts, such as the blades. Floating wind turbines and hybrid marine platforms are the core of the consideration herein. However, the other systems such as wave-energy converters and fixed offshore wind turbines are covered as well.

9.2 Blade Element Theory

The aerodynamic forces consist of the lift and drag forces. The lift forces, skin friction and pressure viscous drags are the main sources of the aerodynamic forces for the slender parts of a wind turbine.

When wind blows, the blades are rotating due to passage of air particles and correspondingly change of air pressure around them. The velocity and pressure are connected; remember the Bernoulli equation and mass conservation principle for an inviscid fluid.

$$P + \frac{1}{2}\rho V^2 = constant$$

$$\rho A V = constant \tag{9.1}$$

© Springer International Publishing Switzerland 2014

M. Karimirad, *Offshore Energy Structures*, DOI 10.1007/978-3-319-12175-8_9

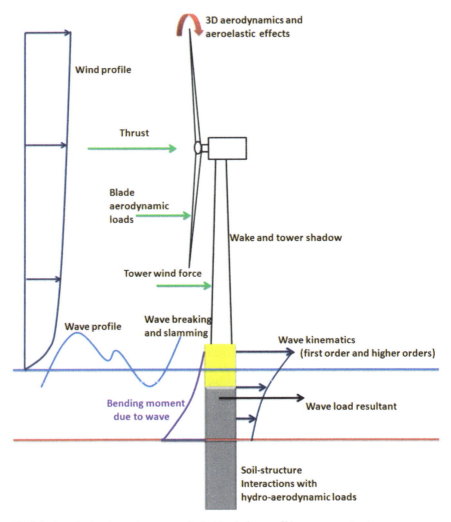

Fig. 9.1 Aero-hydro-dynamic, wave and wind loads for an offshore energy structure

The longer distance the air particle needs to move, the more velocity it needs. In other words, keeping in mind the continuity principle, the flow area is reduced at the top, which results in higher particle velocity. This means that the airfoil shape of the blades makes it possible to have a higher pressure under the blade and get an upward force called 'lift'. The lift force is not the only load present. Putting any object in the passage of fluid disturbs the passage of air particles, and hence resistance appears. This resistance is called 'drag'. The boundary layer is formed around the body, which is a distribution of the fluid particle from zero velocity at the surface of the object to the air velocity at a vertical distance from the surface. Viscous drag or skin friction is linked to roughness. Smoother the surface, less viscous drag appears. Hence, the airfoils are smoothly finished. Pressure drag is due to the wake at the tail of the airfoil. The pressure of the air flow at upstream is higher than the wake pressure in downstream. Both skin friction and pressure-induced drag appear for an airfoil. There are two other forms of drag which affect the airfoil drag: wave drag and induced drag. Wave drag is caused by shock waves over the airfoil that converts energy of the flow to heat and making drag. The induced drag appears for real airfoils in which the length is finite (three-dimensional effects). Figure 9.2 illustrates the nomenclatures of an airfoil. Airfoil is a two-dimensional cross-section of a blade. Camber is the maximum distance between the mean camber line and the chord line. Camber, shape of mean camber line and thickness influence the aerodynamic performance, including the lift, drag and moment characteristics of an airfoil.

Figure 9.3 presents the airfoil aerodynamic forces. Drag force is parallel to the direction of relative air flow, and the lift is perpendicular to relative flow direction. Aircrafts, helicopters, wind turbines, ocean current turbines, hydrofoil marine vehicles and similar structures use the advantage of produced lift for specially-shaped parts, airfoil and hydrofoils.

Consider an airfoil which is not rotating (i.e. for a parked wind turbine, $V = V_{rel}$), the aerodynamic forces can be derived as following if the entire blade is made of a uniform airfoil section.

$$Lift = C_L \left(\frac{1}{2} \rho V^2 \right) S$$

$$Drag = C_D \left(\frac{1}{2} \rho V^2 \right) S$$

$$Moment = C_M \left(\frac{1}{2} \rho V^2 \right) Sc, \tag{9.2}$$

in which ρV^2 is the dynamic pressure, c is the chord length, S is the blade area, C_L is the lift coefficient, C_D is the drag coefficient and C_M is the aerodynamic moment coefficient. S is simply the blade length multiplied to chord dimension. If the blade is made of different airfoil sections, the aerodynamic coefficients are changing along the length and integration over the blade length is needed. For example, the lift can be represented as:

$$Lift = \frac{1}{2} \rho \sum_{i=1}^{N} C_{L_i} \times V_i^2 \times c_i \times \Delta L_i. \tag{9.3}$$

Fig. 9.2 Airfoil nomenclature

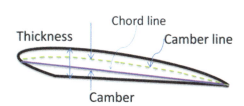

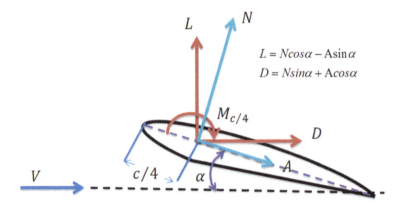

Fig. 9.3 Airfoil aerodynamic forces

The reference area used in drag coefficient depends on the object. For many objects, the reference area is the frontal area, i.e. the cross-sectional area when viewed from ahead. For example, for a wind turbine tower segment, the cross section for drag is $D_i L_i$, in which D_i and L_i are the diameter and length of that segment, respectively. However, for airfoils the reference area is not the frontal area, as it is clear in the lift and drag force defined above. Using the averaged diameter of National Renewable Energy Laboratory (NREL) land-based turbine (Jonkman 2007), if wind shear effect is neglected, one may calculate the drag force on the tower as follows:

$$D_{ave} = \frac{D_{bottom} + D_{top}}{2} = \frac{6 + 3.87}{2} = 4.935m$$

$$R_n = \frac{\rho V_{rated} D_{ave}}{\mu} = \frac{1.22 \times 11.4 \times 4.935}{1.983E-5} = 3.46E6,$$

in which D_{bottom} and D_{top} are the diameter of the tower at the bottom and top, respectively. V_{rated} is the rated-wind speed of the rotor. The projected area of the tower can be assumed to be $S = D_{ave} H$; H is the height of the tower. From Fig. 9.4, the corresponding drag coefficient is (roughly) 0.64 for Reynolds number of 3.5E6.

$$Drag = C_D \left(\frac{1}{2} \rho V^2 \right) S = 0.64 \times (0.5 \times 1.22 \times 11.4^2) \times 4.935 \times 87.6 = 21.93 \text{ kN}$$

The following data in Fig. 9.4 came from Roshko, 'Experiments on the Flow Past a Circular Cylinder at Very High Reynolds Numbers'; refer to Roshko (1960).

Thrust for such a 5 MW turbine is about 750 kN at rated-wind speed. So, for a land-based wind turbine, the drag force on the tower is negligible compared to aerodynamic thrust of the rotor in operational wind conditions. Later, we will see that

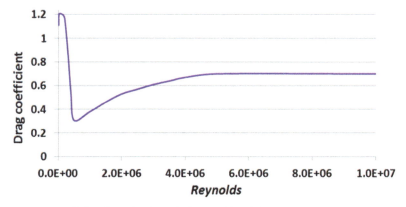

Fig. 9.4 Drag coefficient for a circular cylinder (Roshko 1960)

the tower drag can be important for floating wind turbine when it comes to extreme environmental conditions.

The blades and the tower of wind turbine are slender structures, and hence, the 2D aerodynamic theory is applicable for them. Through the blade element momentum (BEM) theory, the lift and drag coefficients are used to model the aerodynamic forces. If the blades are not rotating, i.e. for a parked wind turbine, the aerodynamic forces are calculated by applying the relative wind speed, while for an operating wind turbine, the induced velocities and wake effects on the velocity seen by the blade elements should be accounted for.

In wind turbines, the span-wise velocity component is much lower than the stream-wise component. Many aerodynamic models assume that the flow at a given point is two-dimensional and the 2D aerofoil data can be applied (Hansen 2008). This simplifies the problem and eases the aero-elastic modeling of wind turbines. The method resembles to 'strip theory', which was widely applied in marine hydrodynamics for the motion analysis of slender ocean structures.

The performance of an airfoil is affected by Reynolds number, surface roughness, Mach number as well as angle of attack. Figure 9.5 illustrates a transversal cut of the blade element. The aerodynamic forces acting on the blade element are shown in this figure. The blade element moves in the airflow at a relative speed V_{rel}. The lift and drag coefficients are defined as follows:

$$C_L(\alpha) = \frac{f_L}{\frac{1}{2}\rho V_{rel}^2 c}$$

$$C_D(\alpha) = \frac{f_D}{\frac{1}{2}\rho V_{rel}^2 c}, \tag{9.4}$$

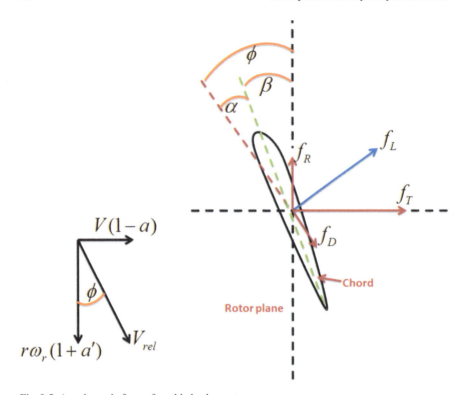

Fig. 9.5 Aerodynamic forces for a blade element

Where f_D and f_L are the drag and lift forces (per length), c is the chord of the airfoil, ρ is the air density, α is the angle of attack and V_{rel} is the relative velocity.

$$V_{rel} = V\sqrt{(1-a)^2 + \left(\frac{r\omega_r}{V}(1+a')\right)^2}$$

$$\alpha = \phi - \beta$$

$$\tan(\phi) = \frac{V}{r\omega_r}\frac{1-a}{1+a'}, \tag{9.5}$$

Where a and a' are the axial and rotational induction factors, respectively, V is the upstream wind velocity, f_T is the thrust force, r is the distance of the airfoil section from the blade root, and ω_r is the rotational velocity (rad/sec). a and a' are functions of ϕ, C_L, C_D and the solidity (fraction of the annular area that is covered by the blade element). The aerodynamic theories to calculate the wind loads for operational and parked conditions are very similar. For a parked wind turbine, the rotational speed (ω_r) is zero as the blades are fixed and cannot rotate. ϕ is 90°, which means the relative wind velocity and the wind velocity are parallel (Karimirad 2011).

An ideal airfoil is relatively insensitive to roughness effects. NREL made an effort to develop special-purpose airfoils for horizontal-axis wind turbines (HAWTs), which in general has a performance requirement that they exhibit a maximum lift coefficient, which is relatively insensitive to roughness effects (Tangler and Somers 1995). For stall-regulated rotors, better peak-power control is achieved through the design of airfoils that restrain the maximum lift coefficient. Restrained maximum lift coefficient allows the use of more swept disc area for a given generator size. Also, for stall-regulated rotors, thicker tip airfoils help to accommodate over-speed control devices. For variable-pitch and variable-rpm rotors, airfoils having a high maximum lift coefficient lend themselves to blades with low solidity. Airfoils having greater thickness result in greater blade stiffness and tower clearance. Airfoils of low thickness result in less drag and are better suited for downwind machines (Tangler and Somers 1995). Recently, in the design of the Technical University of Denmark (DTU) 10-MW reference wind turbine, thick airfoils are used to make it possible for having a light rotor through increasing the structural stiffness of the blades (Bak et al. 2013).

9.3 Aerodynamics of Wind Turbines

Aerodynamic for wind turbines is based on the BEM theory, which is briefly explained in the previous section. The following features need to be included in the aerodynamic model of wind turbines (Burton et al. 2008):

- Deterministic aerodynamic loads: steady (uniform flow), yawed flow, shaft tilt, wind shear, tower shadow and wake effects
- Stochastic aerodynamic forces due to the temporal and spatial fluctuation/variation of wind velocity (turbulence)
- Rotating blades aerodynamics, including induced flows (i.e. modification of the wind field due to the turbine), three-dimensional flow effects and dynamic stall effects
- Dynamic effects from the blades, drive train, generator and tower, including the modification of aerodynamic forces due to vibration and rigid-body motions
- Subsystem dynamic effects (i.e. the yaw system and blade pitch control)
- Control effects during normal operation, start-up and shutdown, including parked conditions

For wind turbine, different types of aerodynamic loads present; this makes aero-elastic analysis of wind turbine demanding. The aerodynamic loads can be divided into different types (Manwell et al. 2006):

- Static loads, such as a steady wind passing a stationary wind turbine.
- Steady loads, such as a steady wind passing a rotating wind turbine.
- Cyclic loads, such as a rotating blade passing a wind shear.
- The wind shear, yaw error, yaw motion and gravity induce cyclic loads.
- Transient loads, such as drivetrain loads due to the application of the brake.
- Gusts, starting, stopping, feathering blades and teetering induce transient loads.

- Impulsive loads, i.e. loads with short duration and significant peak magnitude, such as blades passing a wake of tower for a downwind turbine.
- Stochastic loads, such as turbulent wind forces.
- Turbulence is linked to stochastic loading. Turbulence may have a limited effect on the global responses for floating wind turbines. In such cases, one may investigates fatigue and ultimate limit state analyses based on steady wind modeling with an acceptable accuracy.
- Resonance-induced loads, i.e. excitation forces close to the natural frequencies.
- The structure's eigenfrequencies can be the source of resonance-induced loading.

Wind turbine aerodynamic performance is primarily a function of the steady-state aerodynamics, i.e. the power production of an onshore wind turbine is largely affected by mean wind speed load effects. But, there are a number of important steady-state, quasi-static and dynamic load effects that cause increased loads or decreased power production compared to those estimated from the basic BEM theory. Such effects can especially increase the transient loads. Karimirad listed some of the advanced aerodynamic subjects important for wind turbine functionality and structural integrity (Karimirad 2011):

- Non-ideal steady-state aerodynamic issues

    - Decrease of power due to blade surface roughness (for a damaged blade, up to 40 % less power production).
    - Stall effects on the airfoil lift and drag coefficients.
    - The rotating condition affects the blade aerodynamic performance. The delayed stall in a rotating blade compared to the same blade in a wind tunnel can decrease the wind turbine life.

- Turbine wakes

    - Skewed wake in a downwind turbine.
    - Near and far wakes: The turbulence and vortices generated at the rotor are diffused in the near wake, and the turbulence and velocity profiles in the far wake are more uniformly distributed.
    - Off-axis flows due to yaw error or vertical wind components.

- Unsteady aerodynamic effects

    - Tower shadow (wind speed deficit behind a tower due to tower presence).
    - Dynamic stall, i.e. sudden aerodynamic changes that result in or delay the stall.
    - Dynamic inflow, i.e. changes in rotor operation.
    - Rotational sampling. It is possible to have rapid changes in the flow if the blades rotate faster than the turbulent flow rate.

- Instability

    - Aeroelastic instability can occur whenever any modal response creates accompanying periodic aerodynamic forces that feed the resonant response, for instance the negative damping mechanism. The servo-induced negative

damping can present instabilities for resonant motion of floating wind turbines and hybrid marine platforms.
- The cyclic disturbance of the blade's angles of attack can be the prime mechanism of rotor aeroelastic instability.
- The nacelle/tower instability for the horizontal-axis wind turbine is another
  aeroelastic instability that can occur in the absence of corrective control
  system actions. Whirl mode instability can occur in which the shaft moves
  through a conical locus, with the tower deflecting and the hub moving in a
  nearly circular path.
- The other instability is the yaw angle oscillation of the turbine shaft which
  can occur when the tower axis does not pass through the centre of the nacelle/
  rotor combination (Karimirad 2011).

Figure 9.6 shows the aerodynamic thrust, wind loads on the tower and the shear
forces at the top and bottom of the tower for a land-based 5 MW wind turbine. The
5 MW NREL wind turbine is selected and HAWC2 code is applied to find the loads
and load effects in this case.

As shown in Fig. 9.6, the wind loads become more important for survival cases.
The drag forces are the dominant loads on the tower and on the turbine blades for
harsh conditions. In such conditions, the wind turbine is parked and the blades are
set parallel to the wind in order to minimize the aerodynamic loads on the rotor and
protect the sensitive structural parts, such as blades. The blades are feathered to be
parallel to wind for harsh and storm wind conditions. This reduces the wind loads
and helps the wind turbine to survive during extreme events. Survival load cases
also require considering emergency shutdown incidents.

Two peaks are clear for the loads: one at the rated-wind speed, 11.4 m/sec, and
the other one at harsh conditions, 50 m/sec wind speed. This wind speed is associated with return period of 50 years. The wind speed is referring to the mean wind

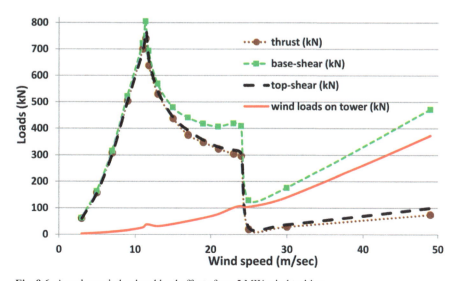

Fig. 9.6 Aerodynamic load and load effects for a 5 MW wind turbine

speed at the top of the tower; and it is 10-min averaged. Figure 9.6 shows the loads due to steady wind (turbulence increases the loads). The tendency in offshore wind engineering is to increase the rated power for each unit. This means larger diameter of the rotor to harness more wind energy. Larger rotor ends up with longer blades and increased hub height. Figure 9.7 illustrates the rated power versus rotor diameter. An extrapolation was applied to estimate a rotor diameter for a 10 MW wind turbine. Such a turbine may have a large diameter in order of 180 m which demands an accurate aerodynamic load calculation accounting for the aeroelasticity, turbulence effects as well as dynamic performance of the support structure in a detailed and integrated format.

9.4   Wind Turbine Aero-Servo Loads

Aero-servo loads on the wind turbines include (a) the direct loads from the inflowing wind, (b) indirect loads that result from the wind-generated motions of the wind turbine, (c) the servo loads in operation of the wind turbine as well as (d) wave-induced aerodynamic loads. Usually, aeroelastic load models are used to determine the aerodynamic wind loads on the rotor and the tower. For offshore wind turbines including floating wind turbines aero-hydro-servo-elastic codes are required to investigate the load and load effects including the aero-servo loads:

a. Direct wind-generated loads:

  - Aerodynamic blade loads during operation, parking, idling, braking and start-up.
  - Aerodynamic drag forces on tower and nacelle.

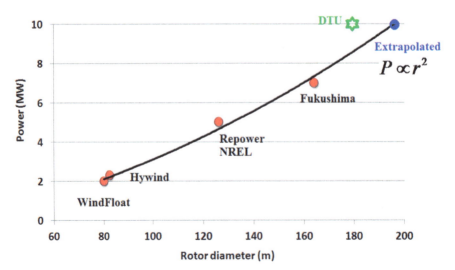

Fig. 9.7 Rated power versus rotor diameter. *NREL* National Renewable Energy Laboratory, *DTU* Technical University of Denmark

b. Indirect loads produced

- Gravitational loads on the turbine components (for blades gravitational loads vary due to rotation).
- Centrifugal and coriolis forces due to rotation.
- Gyroscopic forces due to yawing.
- When the turbine is yawing during the operation, gyroscopic loads on the rotor will occur. This leads to a yaw moment about the vertical axis and a tilt moment about a horizontal axis in the rotor plane.
- Braking forces on the drivetrain (in shutdown events).

c. Servo loads
Control of wind turbine acquires actuators to feather the blades and adjust the yaw of the rotor. The generator torque is controlled as well. All these actions introduce mechanical servo loads to the system.

d. Wave-induced aerodynamics
Wave loads cause the floating wind turbines and hybrid marine platforms to move. These motions affect the relative velocity as well as aerodynamic loads for the turbine.

When determining the aero-servo loads, the following items should be accounted for in the modeling of the loads (DNV 2013):

- Tower shadow and vortex shedding due to the presence of the tower and disturbances of the wind flow when passing the tower; refer to Karimirad (2012)
- Wake effects in wind farms
- Misaligned wind flow, i.e. yaw error and misalignment between wave and wind; refer to Jiang et al. (2012)
- Rotational sampling, for example, due to rotation of blades and their movement through vortices, low-frequent turbulence will be transferred to high-frequent loads
- Aeroelastic effects

  - Possible blade-vibration instabilities caused by stall.

- Turbulence and gusts; refer to Karimirad and Moan (2012)
- Damping

  - Structural damping depends on the blade and tower material.
  - Aerodynamic damping.

- Wind turbine controller effects

  - Aerodynamic imbalance and rotor-mass imbalance as the blades are feathered.
  - Limiting loads through blade pitching.

- Coherence of the wind and the turbulence spectrum of the wind

Different aeroelastic codes have been developed to account for the items mentioned above such as HAWC2, Flex5, Simo-Riflex, Bladed and FAST. These codes have been being under development to carry out analysis needed for offshore wind turbines including floating wind turbines. In recent years, some of these codes applied couple-integrated aero-hydro-servo-elastic time domain dynamic analysis to consider stochastic wave and wind loading on the floating and fixed offshore energy structures.

9.5 Wave Loads and Hydrodynamics

The wind turbine aerodynamic loads occur simultaneously with other environmental loads, such as loads from waves, current and water level. The joint wave and wind loads should be considered for the design of offshore energy structures accounting for wind loads and their companion wave load, current and water level conditions. Herein, the wave loads and hydrodynamics are explained.

Some of the hydrodynamic aspects of offshore energy structures depending on the concept and site specifications are listed below:

- Suitable wave kinematics models
- Hydrodynamic models accounting for water depth, metocean and design/concept specifications
- Extreme hydrodynamic loading including breaking waves
- Nonlinear wave theories and appropriate corrections
- Slamming, ringing and high-order wave loading
- Stochastic hydrodynamics applying linear wave theories with required corrections
- Slender or large-volume structures (and structural components)

Ringing is a transient structural response. When a steep and large wave encounters the structure, high-frequency nonlinear wave loads may excite the eigenmodes of structure-making transient response in the structural response, i.e. in the global-bending moments/effects. Ringing may occur if the lowest structural mode does not exceed three/four times the wave frequency. For more information, refer to Faltinsen (1999).

The relative magnitude of the wave loads can be high for offshore energy structures considering the size and type of the support structure and turbine. Hydro loads may be significant and can be the main cause of fatigue and extreme loads that should be investigated in coupled analysis. Hence, theoretical methods applied for determining the hydrodynamic loads can have an important effect on the cost of the system, structural integrity, reliability, functionality as well as the structure ability to withstand environmental and operational loads.

Offshore renewable energy structures are quite novel and innovative, involving large uncertainties for load and response calculation. Theoretical calculations should be validated against model tests or full-scale measurements. Such experiments (a) confirm that no important hydrodynamic feature has been overlooked, (b) support theoretical calculations and (c) validate theoretical methods. Verification of the implemented theories is required to confirm the accuracy of the coding. Different theories can be compared to advance the code development and perform the verification. Note is required to the fact that validation of a hydrodynamic code is just possible when comparison to experiments are carried out. Meanwhile, both verification and validation of numerical tools handling the hydro loads are needed.

Wave kinematics and wave theories are explained in the previous chapter. To calculate the hydro loads, a recognized wave theory for the representation of the wave kinematics considering the validity range in the specified water depth and metocean conditions is applied. Figure 9.8 shows the relative importance of drag, inertia and diffraction wave forces.

There are several methods proposed for hydro loads including the panel method, Morison formula and pressure integration method or a combination of these methods. The selection of the method should be design/concept dependent; meaning the shape, size as well as the type of the structure are considered when selecting the hydrodynamic method. For slender structures, such as spar, jacket and monopile structures, Morison formula is usually applied to determine the hydro loads while for large-volume structures, as the wave kinematics are disturbed by the presence of the structure, wave diffraction analysis is performed to determine local (pressure) and global wave loads. When the structure moves, i.e. for floating structures, wave radiation forces are important and should be included. Panel methods, i.e. Boundary Element Method (BEM) can be used for analysing the diffraction/radiation

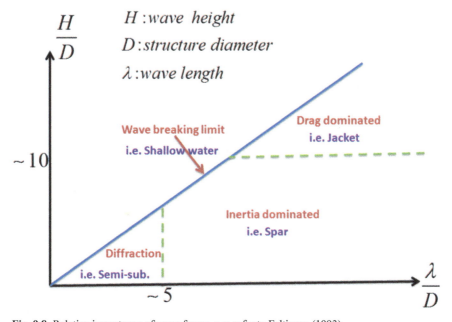

Fig. 9.8 Relative importance of wave forces, e.g. refer to Faltinsen (1993)

problems. BEMs and Boundary Integral Methods (BIMs) are numerical methods which can solve complex engineering problems, including hydrodynamic load calculations. The boundary (not the volume) of domain is discretized into panels on which the related quantities, e.g. velocity potential, are approximated. The velocity potential is considered as a distribution of known shape but of unknown strength; and, unknown strengths are determined by applying the specified boundary conditions through solving an integral equation (e.g. Green's identity) over the boundary.

For offshore energy structures, both viscous and potential flow effects may be important in determining the wave-induced loads. Potential flow is based on inviscid and irrotational assumptions. Hence, viscous effects (i.e. linear and quadratic) should also be included, as the potential flow just considers the wave diffraction/radiation effects.

9.6 Wave Forces on Slender Structures

If diffraction forces are not important compared to other wave forces, refer to Fig. 9.8, the wave forces can be presented by Morison formula. Usually, for slender structures, the diffraction effects are not significant as the structure does not significantly disturb the wave pattern. Hence, the Morison formula can be applied to present wave forces on slender structural members, such as a cylinder submerged in water.

9.6.1 Morison Formula for Fixed Structures

The Morison formula is practical for slender structures where the dimension of the structure is small compared to the wave length, i.e. $D < 0.2\,\lambda$ where D is the characteristic diameter, and λ is the wave length. In other words, it is assumed that the structure does not have a significant effect on the waves. The hydrodynamic forces through the Morison formula include the inertial and quadratic viscous excitation forces. The inertial forces in the Morison formula consist of diffraction and Froude-Krylov (FK) forces for a fixed structure. The horizontal force on a vertical element dz of a cylinder (at level z) is expressed as:

$$dF = dF_m + dF_d = C_m \rho\pi\frac{D^2}{4}\dot{u}_w dz + C_d\,\rho\frac{D}{2}\left|u_w\right|u_w dz, \qquad (9.6)$$

where the first term dF_m is an inertia force and the second term dF_d is a drag force. C_m and C_d are inertia and quadratic drag coefficients, respectively. ρ is the mass density of sea water, D is the cylinder diameter, $\dot{u}_w$ and u_w are the horizontal acceleration and velocity of the water particle velocity, correspondingly. The positive force direction is in the wave propagation direction.

$1.5 \leq C_m \leq 2$ and $0.6 \leq C_d \leq 1.2$ represent good candidates for most of the slender cylindrical structural members in marine structure fields. In general, drag and inertia coefficients are functions of the Reynolds number, Keulegan-Carpenter and the relative roughness. The coefficients also depend on the cross-sectional shape of the structure and of the orientation of the body. For a cylinder with diameter of D, the Reynolds number is defined as $Re = UD / \iota$ and Keulegan-Carpenter number as $KC = UT / D$, where U is the horizontal particle velocity, v is the kinematic viscosity of seawater and T is the period of the waves. Drag and inertia coefficients have different values for the extreme waves that govern the Ultimate Limit State (ULS) and for the moderate waves that govern the Fatigue Limit State (FLS). The marine growth may significantly affect the roughness which should be accounted when deriving the drag and inertia coefficients for life-time calculations.

The resulting force and moment can be derived by integrating over the length of the structure from the seabed to the instantaneous water level. If the force is just integrated from the seabed to the mean water-level surface, then, contributions to the force from the wave crest above the still water level are ignored. This is a minor error when the inertia force is the dominating force component (i.e. for spar platforms). The reason is that the acceleration (and correspondingly the inertia forces) has its maximum at still water level. However, the drag force has its maximum when the crest or trough passes the structure. For drag-dominated structure, i.e. jacket members, a significant error may present by ignoring the contribution from the wave crest (DNV 2013). In Fig. 9.8, the drag and inertia-dominated ranges of wave forces are illustrated based on the simple function of wave height, wave length and characteristic diameter of the structure.

9.6.2   Morison Formula for Floating Structures

The hydrodynamic forces per unit length on the floater based on Morison formula, which was extended to account for the instantaneous position of the structure, can be written as (Karimirad 2011):

$$\frac{dF}{dz} = \frac{\rho}{2} C_d D |u_r| u_r + \rho \frac{\pi D^2}{4} C_m \dot{u}_r + \rho \frac{\pi D^2}{4} \dot{u}_W$$
$$u_r = u_W - u_B \tag{9.7}$$

where $\dot{u}_r$ and u_r are the horizontal relative acceleration and velocity between the water particle velocity u_W and the velocity of the body u_B, respectively. The other parameters are defined in the previous sub-section, 'Morison formula for fixed structures'.

For a floating structure, the added mass forces are included in the Morison formula through relative acceleration and the damping forces appear through the relative velocity. The first term is the quadratic viscous drag force, the second term includes the diffraction and added mass forces, and the third term is the FK force

(FK term). A linear drag term $C_l u$, can be added to the Morison formula as well, where C_l is the linear drag coefficient.

9.6.3 Morison Formula with MacCamy-Fuchs Correction

When the diffraction forces are important and dimension of the structure is large compared with the wave length, i.e. when $D \geq 0.2 \lambda$, Morison formula in its base format is not valid. The inertia force is dominating and should be predicted by diffraction theory. For large-volume structures, the MacCamy-Fuchs correction for the inertia coefficient in some cases may be applied.

For a slender circular cylinder, based on the panel method (i.e. BEM), the added mass coefficient C_a is equal to 1. This corresponds to the diffraction part of the Morison formula. The FK contribution can be found by pressure integration over the circumference and for a cylinder in horizontal direction is equal to 1. Consequently, the inertia coefficient $C_m = 1 + C_a$ for a slender circular cylinder is 2.

The MacCamy-Fuchs presented a solution for corrected inertia term in Morison formula which can be used together with the drag term. The maximum horizontal inertia force on a vertical cylinder installed in water depth of h, for linear waves accounting for diffraction, is obtainable as:

$$F_{max,\,horizontal} = \frac{4\rho g}{k^2} \zeta_a \frac{\sinh \left[k(h + \zeta_a \sin \alpha) \right]}{\cosh kh} \xi, \qquad (9.8)$$

where, k is the wave number and ζ_a is the linear wave amplitude. α and ξ are the functions of k and cylinder radius $R_{cylinder}$. Tables defining the related values of α and ξ can be found in the literatures (i.e. DNV 2013). The above formula is just valid for vertical circular cylinder and for other shapes, i.e. when a conical component appears in the support structure, the resulting force (and moment) is different from what is presented herein.

9.6.4 Pressure Integration Method

The pressure integration method consists of integrating the static and dynamic pressures over the wetted surface of the body. The static pressure corresponds to the buoyancy, and the dynamic pressure of the waves corresponds to the wave FK force. The transversal component of the dynamic pressure integration corresponds to the FK term in the Morison formula.

The Morison formula combined with the pressure integration method is a practical approach to model the hydrodynamic forces of slender offshore energy structures. Two approaches combining the pressure integration and Morison formula are:

a. Pressure integration method accounts for the FK part while the diffraction part comes through the Morison formula. Therefore, in such case, the FK term should be removed from the Morison formula (Karimirad et al. 2011).
b. Also, in another approach, it is possible to use pressure integration method for calculating just the vertical forces and forces on conical sections while Morison formula applied for transversal loads (Karimirad et al. 2011).

9.7 Breaking Wave Loads

Breaking waves apply high-impact forces in short duration and consequently affect the performance and fatigue life of the marine structures. As it is mentioned in the previous chapter, in deep water, waves break when $H/\lambda > 0.14$ and in shallow water, when $H/d > 0.78$ (DNV 2007). In shallow water, the breaking wave forces may yield the maximum hydrodynamic loads on structure (in particular, plunging-breaking waves). Waves start to break when they become unstable and dissipate the energy in the form of turbulence. During the wave breaking, the energy is focused close to the wave-crest and wave-energy spreads (Faltinsen 1993).

Based on Stokes criterion for wave breaking, the wave-crest velocity reaches the celerity. Wave breaking depends on several parameters, including water depth, wave height, seabed slope, wave period and steepness. Wave loads from breaking waves depend on the type of breaking waves, i.e. surging, plunging and spilling waves. Based on the type of breaking waves, the wave kinematics and consequently the wave loads are different.

Quasi-static model may be applied to present the spilling and surging wave forces. For bottom-fixed offshore wind turbines, spilling and plunging are the most relevant. The energy of plunging-breaking waves is dissipated over a small area with high impulsive loads and pressures. Wind–wave, wave–wave and wave–current interactions affect the breaking wave properties. The uncertainties in breaking wave forces are mainly due to the flow kinematics and the relationship between flow and forces (Chella et al 2012). The breaking wave forces can be defined as impact (similar to slamming) loads. The impact force will be added to inertia and drag forces in the Morison formula and the total load on the structure is: $F = F_{inertia} + F_{drag} + F_{impact}$

The impact force from a plunging wave can be expressed as:

$$F_{impact} = \frac{1}{2}\rho A C_{S} u_r^2, \tag{9.9}$$

in which u_r is the wave-crest celerity (relative to structure), A is the area exposed to the slamming force and C_S is the slamming coefficient. Coefficients between π and 2π are good representatives for slamming coefficient of circular cylinders. Careful selection of slamming coefficients for structures should be made based on standard requirements. Slamming area depends on different items including:

- How much of the wave crest is active during impact?
- How far has the plunging breaker come relative to the structure?
- How wide or pointed is the breaker (when it hits the structure)?

The associated wave forces of surging and spilling breakers on a vertical cylindrical structure of diameter D is represented in (DNV 2013). Based on this approach, the cylinder is divided into a number of sections. The instantaneous force dF_{impact} per vertical length unit on this section together with underlying sections (which have not yet fully penetrated the sloping water surface) is calculated. When the breaking wave approaches the structure, the time instant when a section is hit by the wave (and starts to penetrate the sloping water surface) is defined using the instantaneous wave elevation close to the cylinder.

$$dF_{impact} = \frac{1}{2} \rho D C_S u_r^2$$

$$C_S = 5.15 \left(\frac{D}{D+19S} + \frac{0.107S}{D} \right)$$

$$0 < S < D \tag{9.10}$$

The penetration distance S for a section is the horizontal distance from the edge on the wet-side of the cylinder to the sloping water surface. S is measured in the direction of the wave propagation (DNV 2007). For fully submerged sections, the wave forces can be determined from the Morison theory explained above. The wave kinematics and water particle velocity are calculated considering the type of breaking wave at the offshore site (Chella et al. 2012).

9.8   Large-Volume Structures

Hydrostatic, static stability, hydrodynamic and wave loads for large-volume marine structures are explained herein.

9.8.1   Hydrostatic Considerations

Stability of the system should be checked in the first place. The bottom-fixed structures are supported by the foundation. The foundation-soil interaction loads need to be balanced with inertia and weight loads. For floating offshore structures, adequate hydrostatic stability should be confirmed; the structural weight, mooring line tension and buoyancy forces should be balanced. Risers and mooring mass and pretensions are part of this load balance. The entire system mass including the support

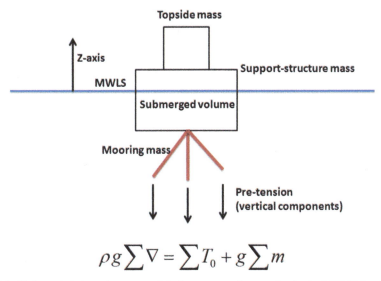

Fig. 9.9 Hydrostatic balance between weight, buoyancy and pretension loads. *MWLS* Mean Water Level Surface

structure, topside, mooring lines and tendons should be accounted for. Mooring line pretension, especially in the case of tension-leg platforms (TLPs), is included in such hydrostatic stability calculations (Fig. 9.9).

The static balance is important for the success of subsequent hydrodynamic analyses. Submerged volume (and consequently buoyancy of large-volume structures) is obtained directly from the wet surface of described geometry defined in the radiation/diffraction analysis. If a dual model, including Morison elements, is applied, the actual location and dimensions of the Morison elements should be considered in a proper manner.

Hydrostatic data can be expressed in the form of surface integrals over the mean body-wetted surface (S) based on the Gauss divergence theorem (WAMIT 2013). The volume of the submerged part (∇) and center of buoyancy (COB) (x_b, y_b, z_b) are defined as:

$$\nabla = -\iint_S n_1 x\, ds = -\iint_S n_2 y\, ds = -\iint_S n_3 z\, ds$$

$$x_b = \frac{-1}{2\nabla} \iint_S n_1 x^2\, ds$$

$$y_b = \frac{-1}{2\nabla} \iint_S n_2 y^2\, ds$$

$$z_b = \frac{-1}{2\nabla} \iint_S n_3 z^2\, ds \qquad\qquad (9.11)$$

n_1, n_2, n_3 are normal vectors in x, y and z directions; the z-axis is upward. Matrix of hydrostatic and gravitational stiffness (restoring) is defined as follows:

$$C(3,3) = \rho g \iint_S n_3 ds$$

$$C(3,4) = \rho g \iint_S y n_3 ds$$

$$C(3,5) = -\rho g \iint_S x n_3 ds$$

$$C(4,4) = \rho g \iint_S y^2 n_3 ds + \rho g \nabla z_b - mgz_g$$

$$C(4,5) = -\rho g \iint_S xy n_3 ds$$

$$C(4,6) = -\rho g \nabla x_b + mgx_g$$

$$C(5,5) = \rho g \iint_S x^2 n_3 ds + \rho g \nabla z_b - mgz_g$$

$$C(5,6) = -\rho g \nabla y_b + mgy_g. \tag{9.12}$$

In the above equations, x_g, y_g, z_g is referring to the center of gravity (COG) of the structure and m denotes the body mass. The indices 1, 2, 3, 4, 5 and 6 are referring to linear and angular motions of the platform: surge, sway, heave, roll, pitch and yaw, correspondingly. $C(i, j)$ refers to restoring load (force or moment) for i-motions due to j-motions; $C(i, j) = C(j, i)$ except for $C(4, 6), C(5, 6)$. The other values of the matrix are zero, especially: $C(6,4) = C(6,5) = 0$. For free-floating structures, the buoyancy and weight are in balance: $m = \rho \nabla$. Otherwise, for structures which are taut (i.e. TLPs), care is needed. WAMIT allows users to define an alternative form of inputs to describe the total mass of the system (WAMIT 2013). For a freely floating body and structures in which the difference between the weight and buoyancy force is negligible, i.e. for semisubmersible and spar platforms, equilibrium of static forces require that the center of gravity and the center of buoyancy must lie on the same vertical line, which result in $x_g = x_b, y_g = y_b$, and consequently, $C(4,6) = C(5,6) = 0$. Normally, the origin of the coordinate system is chosen in a way that $x_g = y_g = 0$ (i.e. at the center of gravity or at mean water-level surface).

Figure 9.10 shows a free-floating structure and the metacentric height for heeling. The relation of the metacentric height (GM) with the center of mass and center of buoyancy is illustrated in this figure. Similar figure and relation can be derived for tilting. Traditionally, in naval architecture, heeling and tilting are used for static or mean value of roll and pitch motions.

$$GM = KM - KG = BM + KB - KG$$

$$GM_T = \frac{\iint_S y^2 n_3 ds}{\nabla} + KB - KG$$

$$GM_L = \frac{\iint_S x^2 n_3 ds}{\nabla} + KB - KG$$

$$\tag{9.13}$$

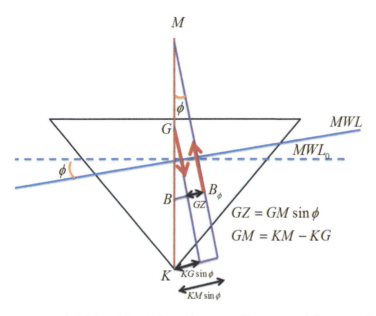

Fig. 9.10 Metacentric height and its relation to the center of buoyancy and the center of gravity, in a stable system the interaction of the buoyancy and weight provides enough righting moment against external heeling (or tilting) moments. *MWL* Mean Water Level

The hydrostatic stiffness in roll and pitch for a freely floating structure are given by:

$$C(4,4) = \rho g GM_T$$
$$C(5,5) = \rho g GM_L. \tag{9.14}$$

For platforms with concentrated small water plane area such as spar, the effect of area moment of inertia is negligible compared to weight-buoyancy effects. Hence, the spar-type platform is ballast stabilized.

$$C(4,4) = C(5,5) = \rho g GM \simeq \rho g(KB - KG) \tag{9.15}$$

This means the center of gravity of such structures (i.e. spars) should be set below the center of buoyancy to provide enough stability in roll and pitch motions. The ship-shaped platforms, barge and semisubmersibles get their stability from area moment of inertia rather than the mass-buoyancy contribution. Hence, these structures are surface-area stabilized. For ships and barge a large area is needed while for semisubmersibles the surface-area is distributed from center of area to provide larger righting arm.

Applying the correct center of gravity in the hydrodynamic analyses is important. Also, the center of buoyancy, submerged volume and metacentric height (for free-floating structures) should be carefully checked. The following issues must be noticed while determining the metacentric height and restoring matrix:

1. Influence from free surface effects in internal tanks
2. The additional restoring effects due to the reaction from buoyancy cans

3. Closed cushions and additional restoring from them, i.e. for heave restoring
4. Stiffness contributions from tethers, mooring lines, risers and possible additional restoring from thrusters

For taut-moored structures, in particular for TLPs, the effect of tension in stiffness and stability is considerable and the stability of such structures is highly influenced by tension magnitude and its variations in tendons (Karimirad et al. 2011).

9.8.2   Mass and Inertia Loads

The mass distribution of the system may either be entered as a global mass matrix (for rigid body dynamics) by defining the mass and mass moments of inertia, or by a detailed mass distribution, i.e. finite element modeling. Note that the coordinate system may be referred to the centre of gravity or the mean water plane. For a free-floating structure (stable body without external constraints), as it is mentioned above: $m = \rho V$ and $x_g = x_b$, $y_g = y_b$. Also, the following expressions stand for the mass matrix:

$$
M = \begin{pmatrix}
m & 0 & 0 & 0 & mz_g & -my_g \\
0 & m & 0 & -mz_g & 0 & mx_g \\
0 & 0 & m & my_g & -mx_g & 0 \\
0 & mz_g & my_g & I_{11} & I_{12} & I_{13} \\
mz_g & 0 & -mx_g & I_{21} & I_{22} & I_{23} \\
-my_g & mx_g & 0 & I_{31} & I_{32} & I_{33}
\end{pmatrix}.
\tag{9.16}
$$

The mass moment of inertia is defined as:

$$
I_{ij} = \int\limits_{structure} (r^2 \delta_{ij} - x_i x_j)dm
$$
$$
r^2 = x_1^2 + x_2^2 + x_3^2,
\tag{9.17}
$$

where Kronecker delta symbol is: $\delta_{ij} = \begin{cases} 0 & i \neq j \\ 1 & i = j \end{cases}$.

As an example:

$$
I_{11} = \int\limits_{structure} \left(x_2^2 + x_3^2\right)dm.
\tag{9.18}
$$

The mass moment of inertia may be defined using the following expression as well:

$$
I_{ij} = mr_{ij}\left|r_{ij}\right|.
\tag{9.19}
$$

The parallel axis theorem is used to calculate mass moment of inertia in a desired coordinate system using the moments of inertia defined in another coordinate system. For a free-floating structure, the pitch/roll mass moment of inertia in coordinate system passing the mean water-level surface (I_{MWL}) and pitch/roll mass moment of inertia in coordinate system passing center of gravity (I_{COG}) has the following relation:

$$I_{MWL} = I_{COG} + mz_g^2. \tag{9.20}$$

9.8.3 Hydrodynamic Considerations

As mentioned before, Morison formula is usually applied for slender marine structures to calculate the wave hydrodynamic loads. There are some modifications proposed for the Morison formula to account more accurately the diffraction effects when the structure dimensions increase compared to wave length (i.e. MacCamy-Fuchs approaches). However, when the structure compared to waves is large (i.e. $D > 0.2 \lambda$), the effects of the structure on wave and diffraction effects become more important. Hence, the Morison formula is not accurate enough to model hydrodynamics of such structures and diffraction should properly be considered.

A large-volume structure can be bottom-fixed such as gravity based structures or it can be floating, such as ships, spar, TLP and semisubmersible platforms. The focus of this section is rigid-body associated hydrodynamics based on main references on wave-induced loads/load effects of large-volume structures (e,g.Newman 1977; Faltinsen 1993).

Linear superposition of regular wave components can be applied to obtain the wave loads in irregular sea. Analysing a structure in regular incident wave is so-called a 'frequency domain analysis'. In frequency domain analysis, it is assumed that the load and responses are steady state (all transient effects are neglected), and they are harmonically oscillating with the same frequency as the incident waves. In the case of a forward speed (for example ships), the loads and responses are oscillating with the encounter frequency.

The linear hydrodynamic analysis in frequency domain consists of two main parts (e.g. refer to Faltinsen 1993).

1. Radiation problem

If a structure is forced to oscillate in calm water with the wave frequency (assuming rigid body motions without incident waves), the structure generates waves. This is called radiation problem. In this condition, the loads applied on the structure consist of added mass, damping and restoring loads.

$$F_k^{Radiation} = -A_{kj} \frac{\partial^2 \eta_j}{\partial t^2} - B_{kj} \frac{\partial \eta_j}{\partial t} - C_{kj} \eta_j, \tag{9.21}$$

where A_{kj} and B_{kj} are added mass and potential damping which are functions of the wave frequency, and, C_{kj} are the hydrostatic restoring coefficients. $j, k = 1, 2, \ldots 6$ are indices for the six degrees of rigid body motions.

2. Diffraction problem

If the structure is restrained from motions and is encountered by incident waves, the structure resists against the applied wave-induced loads. The resulting wave excitation loads are:

$$F_k^{Diffraction} = f_k(\omega) \exp(-i\omega t)$$
$$k = 1, 2, \ldots 6.$$

(9.22)

The excitation loads consist of diffraction and FK forces/moments. The part of the wave excitation loads that is given by the undisturbed pressure in the incoming waves is called FK loads (DNV 2007).

9.8.4 Hydrodynamic Analyses Methods

Potential theory can be used to assess the wave-induced loads on large-volume structures. In this method, hydro loads are obtained from a velocity potential of the irrotational fluid motion assuming an incompressible and inviscid fluid. The BEM is the most common numerical method for solving the potential flow. The velocity potential in the water is characterized by sources over the mean wet structure surfaces. The source function satisfies the free-surface condition, so-called free-surface Green function. Source strength is found from integral equations by satisfying the boundary condition on the body surface (Lee 1995). An alternative is Rankine source method. Rankine sources (1/R) are distributed over both the mean wetted surface and the mean free surface (Rankine source method is preferred for forward speed problems, i.e. in ship hydrodynamics). In such method, the mean wet surface is discretized with panels (so, the method is called 'panel methods').

Several wave periods and headings should be included in the analyses such that the dynamic motions and forces/moments can be predicted with acceptable accuracy. For a linear motion analysis in the frequency domain, computations are normally performed for 30–40 frequencies. When a resonance peak is close to the wave spectral frequency range, more frequencies may apply to get accurate responses. For example, for a semi-submersible wind turbine, the linear wave frequency domain analysis based on panel method considering frequency step of 0.05 covering 0.0–2.0 rad/sec range (with smallest wave period of 3.14 s) means 41 frequencies are involved. This makes analysis time consuming depending on the applied panel method and panel size. A low-order panel method uses flat panels while a higher order panel method uses curved panels. A low-order method is more time demanding relative to a higher-order one (WAMIT 2013). Simulation time rapidly grows by increasing the number of panels; hence, the minimum adequate number of panels depending on the required level of accuracy is usually applied. There are some proposed requirements and recommendations for panel (mesh) size in order to get better results. Moreover, sensitivity studies and convergence tests are applied to support the analysis accuracy.

Panel mesh affects the results. Modeling principles provided in literatures are important to apply proper mesh. For a low-order panel method (BEM) with constant value of the potential over the panel, the following principles are given in DNV standards (e.g. refer to DNV 2007):

- Panel mesh diagonal length should be less than 1/6 of smallest wave length considered.
- Finer mesh should be used in areas with sudden changes of geometry, i.e. edges/ corners.
- Finer mesh should be used towards the water surface.
- Water plane area and the volume of the discretized-model should be close to real structure data.
- For calculating wave-surface elevation and fluid-particle velocities, diagonal of a typical panel should be less than 1/10 of the shortest wave length considered.

Finite element method (FEM) can be used to solve the potential flow problems by discretizing the volume of the fluid domain by the elements. Also, semi-analytic expressions can be derived for the solution of the potential flow problem of simple geometries (like sphere, cylinder and torus). For certain fixed or floating offshore structures with simple geometries, like Spar platform, such solutions can be useful. Another approach to calculate wave-induced loads is the strip theory. Hydro loads on slender large-volume offshore energy structures may be predicted by strip theory where loads are obtained by summation of loads on two-dimensional strips.

Hydrodynamic damping of large-volume offshore structures is due to (1) wave radiation damping, (2) hull-skin friction damping, (3) hull-eddy making damping, (4) viscous damping from bilge keels and other appendices as well as (5) viscous damping from risers and mooring. Potential theory accounts for the wave-radiation damping (potential damping). Simplified hydrodynamic models, experiments and computational fluid dynamics (CFD) can be implemented to evaluate viscous damping.

9.8.5 First-Order Wave Loads

Large-volume structures are inertia-dominated and the wave diffraction loads are larger than the drag-induced loads. However, slender members require a Morison load model to account for the drag terms. For example, for spar and semisubmersible, in addition to the radiation/diffraction load models, which consider the inertia terms, the Morison elements accounting for the viscous drag loads are required (Karimirad and Moan 2010). To perform global dynamic analysis, the wave frequency-associated loads are usually sufficient to accurately represent the main responses. Linear-wave analyses are widely applied to predict wave-induced responses of marine structures. In a linear-wave analysis, the fluid dynamic pressure and the wave-induced loads are proportional to the wave amplitude. Hence, the loads from individual waves in an irregular/stochastic environmental condition can be superimposed. In the linear wave analysis, just the wetted surfaces up to the mean water-level surface are considered. The following parameters among others are the main output from the linear-wave analysis: (1) hydrostatic, (2) excitation forces, (3) potential damping, (4) added mass, (5) first-order motions, RAOs (response amplitude operator) and (6) Also, it is possible to calculate the mean drift forces/moments from linear analysis (the mean wave-drift force/moments are second order) (e.g. refer to Faltinsen 1993).

(1) Hydrostatic considerations are explained before. (2) The excitation forces in the linear wave analysis can be obtained from Haskind or direct integration methods:

1. Exciting forces from the Haskind relations

$$F_k = -i\omega\rho \iint_S \left(n_k \varphi_0 - \varphi_k \frac{\partial \varphi_0}{\partial n} \right) ds, \tag{9.23}$$

where F_k is the exciting force, and it is the function of wave frequency (ω). $k=1,2,\ldots6$ are indices for the six degrees of rigid body motions. φ_0 is the incident-wave potential. Haskind relation is useful to define the exciting loads when the detailed pressure distribution is not needed. The FK component is defined as the contribution from the incident-wave potential (φ_0). Using the Haskind relations, the FK components correspond to the contributions from the first part. The scattering component is the remainder, second terms in parenthesis of the Haskind equation.

2. Exciting forces from direct integration of hydrodynamic pressure

$$F_k = -i\omega\rho \iint_S (n_k \varphi_D) ds \tag{9.23}$$

By using the direct integration method, FK and scattering terms correspond to the components of the total diffraction potential (φ_D) (refer to Lee 1995 and WAMIT 2013).

(3) and (4) are potential damping and added mass: Due to forced harmonic oscillations of body, added mass and damping forces/moments occur. The surrounding fluid oscillates and creates a pressure field which introduces hydro loads because of these oscillations. Note: added mass is not an amount of water that oscillates with the structure. Added mass and potential damping is hydrodynamic loads coming from pressure fields in water.

$$A_{kj} - \frac{i}{\omega} B_{kj} = \rho \iint_S (n_k \varphi_j) ds \tag{9.24}$$

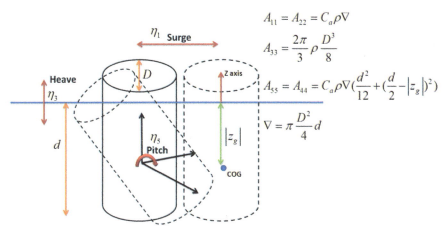

$$A_{11} = A_{22} = C_a \rho \nabla$$

$$A_{33} = \frac{2\pi}{3} \rho \frac{D^3}{8}$$

$$A_{55} = A_{44} = C_a \rho \nabla \left(\frac{d^2}{12} + \left(\frac{d}{2} - |z_g| \right)^2 \right)$$

$$\nabla = \pi \frac{D^2}{4} d$$

Fig. 9.11 Added mass for a circular cylinder

Added mass force is in the form of $-A_{kj} \ddot{\eta}_k$, and potential damping force is represented as $-B_{kj} \dot{\eta}_k$. The added mass (A_{kj}) and potential damping (B_{kj}) are the functions of wave frequency. Also, the added mass and potential damping forces are dependent on the size and shape of the floating part. Figure 9.11 shows added mass for a circular cylinder. For a circular cylinder, the surge/sway-added mass coefficient (C_a) is almost frequency-independent, and it is equal to 1.0. Also, the heave-added mass may be estimated using the disc approach.

Strip theory may be used to calculate the added mass of structures (A_{kj}) using added mass of 2D sections (a_{kj}) and integrating over the length. As an example, some of the relations are listed below:

Most of offshore structures are made of simple geometries, such as circular and rectangular cylinders. As mentioned above, strip theory may be used to integrate added mass of sections along the length of components; and afterward, sum up added mass of components to find out added mass of the system. Added mass for some simple 2D geometries are shown in Fig. 9.13. For more information refer to DNV (2011).

(5) First-order forces and response amplitude operators (RAOs): The next chapter is dedicated to dynamic response analysis. (6) Drift forces: Mean drift forces can be calculated by the pressure integration and momentum methods. The second-order forces are discussed in the following section.

9.8.6 Second-Order Wave Loads

Second-order wave loads, forces and moments, are proportional to the second-order wave amplitude. This includes mean drift, difference and sum-frequency loads. Low-frequency motions of a moored floating offshore structure are caused by slowly varying wave, wind and current forces. Here, the second-order wave

Table 9.1 Strip theory applied for added mass, see Fig. 9.12.

	Surge	Sway	Heave	Roll	Pitch	Yaw
Surge						
Sway		$A_{22} = \int_L a_{22}dx$	$A_{23} = \int_L a_{23}dx$	$A_{24} = \int_L a_{24}dx$	$A_{25} = -\int_L a_{23}xdx$	$A_{26} = \int_L a_{22}xdx$
Heave			$A_{33} = \int_L a_{33}dx$	$A_{34} = \int_L a_{34}dx$	$A_{35} = -\int_L a_{33}xdx$	$A_{36} = \int_L a_{32}xdx$
Roll				$A_{44} = \int_L a_{44}dx$	$A_{45} = -\int_L a_{34}xdx$	$A_{46} = \int_L a_{24}xdx$
Pitch					$A_{55} = \int_L a_{33}x^2dx$	$A_{56} = -\int_L a_{32}x^2dx$
Yaw						$A_{66} = \int_L a_{22}x^2dx$

Fig. 9.12 Strip theory applied for added mass, see Table 9.1

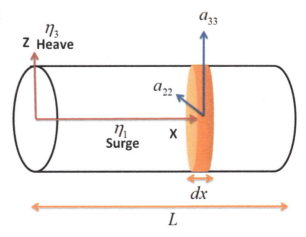

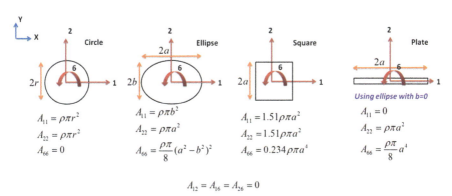

$$A_{12} = A_{16} = A_{26} = 0$$

Fig. 9.13 Added mass for simple 2D geometries, refer to Korotkin (2009)

loads including slowly varying loads (mean drift and slow drift) and high-frequency loads are explained.

To illustrate the second-order wave loads, one may consider the quadratic velocity term in the Bernoulli equation for the fluid pressure $(0.5\rho V^2)$. If we assume a sea state consisting of two linear waves $\zeta = \zeta_{a1} \sin(\omega_1 t + \chi_1) + \zeta_{a2} \sin(\omega_2 t + \chi_2)$, the velocity can be written in the form of $V = V_1 \cos(\omega_1 t + \varepsilon_1) + V_2 \cos(\omega_2 t + \varepsilon_2)$. Hence, the quadratic velocity presents as:

$$
\begin{aligned}
V^2 &= \left[V_1 \cos(\omega_1 t + \varepsilon_1) + V_2 \cos(\omega_2 t + \varepsilon_2) \right]^2 \\
&= V_1^2 \cos^2(\omega_1 t + \varepsilon_1) + V_2^2 \cos^2(\omega_2 t + \varepsilon_2) + 2V_1 \cos(\omega_1 t + \varepsilon_1) \times V_2 \cos(\omega_2 t + \varepsilon_2) \\
&= \frac{V_1^2}{2} + \frac{V_2^2}{2} + \frac{V_1^2}{2} \cos(2\omega_1 t + 2\varepsilon_1) + \frac{V_2^2}{2} \cos(2\omega_2 t + 2\varepsilon_2) \\
&\quad + V_1 V_2 \cos\left[(\omega_2 - \omega_1)t + \varepsilon_2 - \varepsilon_1)\right] + V_1 V_2 \cos\left[(\omega_2 + \omega_1)t + \varepsilon_2 + \varepsilon_1)\right].
\end{aligned}
$$

$$(9.26)$$

- $\dfrac{V_1^2}{2} + \dfrac{V_2^2}{2}$: represents the mean drift forces,
- $V_1 V_2 \cos\left[(\omega_2 - \omega_1)t + \varepsilon_2 - \varepsilon_1)\right]$: represents the slowly varying drift (difference frequency) loads
- $\dfrac{V_1^2}{2}\cos(2\omega_1 t + 2\varepsilon_1) + \dfrac{V_2^2}{2}\cos(2\omega_2 t + 2\varepsilon_2) + V_1 V_2 \cos\left[(\omega_2 + \omega_1)t + \varepsilon_2 + \varepsilon_1)\right]$: represents the high-frequency (sum-frequency) loads

1. Mean drift loads

The wave-induced drift loads have inviscid and viscous parts. The inviscid wave-drift load is a second-order wave force, which is proportional to the square of the wave amplitude. In the design of mooring systems for offshore structures, loads due to current, wind, wave-drift forces and wind- and wave-induced motions are generally of equal importance. Zhao and Faltinsen found the ratio of drift and linear wave loads for a hemisphere floating in the free surface with a full-scale diameter of 50 m to be $\zeta_a / 100$. Hence, for regular waves with amplitude of 1 m, the drift loads are 100 times smaller than the linear wave loads (Faltinsen 1993).

For a spar-floating wind turbine, Karimirad showed that the mean drift forces are almost 1–5 % of the linear wave loads and may be neglected when considering the global dynamic responses. The hydrodynamic drift and second-order forces do not significantly affect the motion and tension responses for such slender structure. However, the heave motion is more affected by the drift and second-order forces. The second-order forces for conventional spar (oil platform) can be important while for a floating spar-type wind turbine the drift is dominated by wind loads and second-order loads are small due to relative slender water surface area compared to probable wave length. Moreover, Karimirad showed that the elastic modes of the mooring lines can be excited by such second-order loads on the platform, which highlight the importance of the mooring line viscous damping in order to damp the resonant responses of the mooring system (Karimirad 2013, Modeling aspects of a floating wind turbine for coupled wave–wind-induced dynamic analyses, Volume 53).

The second-order potential has the same time dependency as the boundary value conditions. The second-order problem involves solving a boundary value problem where the boundary conditions are from the product of two terms. Each term is first order and harmonically oscillates with frequency of linear wave. The product of the terms gives one term that is time independent, and the other term which oscillates with frequency of 2ω. So, second-order velocity potential $\left(\phi_2 \propto \zeta_a^2\right)$ can be written as $A + B\cos(2\omega t + \varepsilon)$, where, A and B are independent of time. The pressure associated with this second-order potential obtained from Bernoulli equation is:

$$-\rho\partial\phi/\partial t = 2\rho\omega B \sin(2\omega t + \varepsilon). \tag{9.27}$$

So, the mean value of pressure part is zero. This means the second-order potential does not result in drift loads. Hence, when calculating the mean wave loads on a structure, it is not necessary to solve the second-order problem and the linear first-order solution can be used. Moreover, drift loads in an irregular sea are obtained by adding results from regular waves.

Two methods are available for computing the wave-drift forces: (1) the near-field method based on the direct pressure integration and (2) the far-field method based on the momentum-conservation principle. Maruo presented the drift surge and sway loads using the conservation of momentum as following (Maruo 1960):

$$\overline{F_i} = - \overline{\iint\limits_{S_\infty} (pn_i + \rho V_i V_n)ds} \quad i = 1, 2, \tag{9.28}$$

where p is the fluid pressure, n_i is a normal vector in surge or sway direction, V_i is the fluid velocity in surge or sway direction, $V_n = \partial \phi / \partial n$ is the normal component of the fluid velocity at the surface. S_∞ is a non-moving circular cylinder away from the body (the method is so-called far filed). Newman derived a similar formula for the mean wave-drift yaw moment (Faltinsen 1993).

Maruo (1960) has used the above equation to present a formula for drift loads of a 2D structure (floating or fixed) subjected to regular incident wave in deep water areas. In such case, three velocity potentials present: (a) Incident wave-velocity potential, (b) reflected wave velocity potential and (c) transmitted wave velocity potential. Transmitted waves are the combination of incident wave and waves generated by the structure (consider a floating structure and its radiation).

By integrating $\overline{F_i} = - \overline{\iint\limits_{S_\infty} (pn_i + \rho V_i V_n)ds}$ $i = 1, 2$ over the 2D surfaces, it is possible to derive the horizontal drift force as $\overline{F_{drift}} = \dfrac{\rho g}{2} A_R^2$. The A_R is the amplitude of the reflected waves. The drift force is in the direction of the propagating waves. The reflected waves (A_R) are linked to diffraction and radiation waves. If the body is good in making waves, then, the drift forces increase. For slender structures compared to the wave length (i.e. spar), the drift forces are negligible. However, for large structures, i.e. ship-shaped offshore structures, the wave drift are becoming important as the surface area of the structure is relatively high compared to wave length even in long waves. If the entire propagating wave is reflected, the maximum drift force happens: $\overline{F_{drift}}\big|_{max} = \dfrac{\rho g}{2} \zeta_a^2$. There are similar formulas for drift force on 3D structures in incident regular waves.

The second approach to derive mean drift loads is the near-field method based on the direct pressure integration method. Applying the Bernoulli equation for the pressure and writing forces/moments on the hull correctly to the second order in wave amplitude leads to mean wave-drift loads. All load components can be obtained by this method. However, in far-field method, just surge, sway and yaw drift loads could be obtained.

Faltinsen has studied the pressure integration method to derive drift loads on a vertical wall (Faltinsen 1993). Let us review this example. The propagating waves are totally reflected from the wall and standing waves form; $\zeta = 2\zeta_a \sin \omega t$ is the standing wave amplitude at the wall. The linear velocity potential can be found as

$\phi_1 = \dfrac{2g\zeta_a}{\omega} \exp(kz)\cos \omega t \cos kx$, where x is the wave propagation direction and wall is located at $x = 0$. Linear wave loads can be calculated by the integration of the pressure up to the mean water-level surface.

$$F_{linear} = -\int\limits_{-\infty}^{0} \rho \dfrac{\partial \phi_1}{\partial t}\bigg|_{x=0} dz = \dfrac{2\rho g \zeta_a}{k}\sin \omega t, \qquad (9.29)$$

where $\overline{F_{linear}} = 0$, as the linear force is oscillating harmonically. Hence, to get mean drift forces, higher-order terms including the integration up to the instantaneous wave elevation should be added (i.e. complete Bernoulli together with solving the hydrodynamic problem to the second order). However, it has been shown that the second-order velocity potential provides zero mean drift loads.

Maruo formula results in drift force of $\dfrac{\rho g}{2} \zeta_a^2$ for a wall, which can be extended to any structure as long as the structure has vertical sides at the mean water level. The total drift force/moment is a function of non-shadow part of the water plane, the shape of the curve at water plane and the wave-propagation direction.

2. Difference frequency loads

The difference frequency (slowly varying drift) loads in an irregular sea state consisting of the N regular wave with a frequency of ω_i $i = 1 : N$ oscillates at difference frequencies $\omega_j - \omega_i$:

$$F^{diff}(t) = \text{Re} \sum_{j=1}^{N}\sum_{i=1}^{N} \zeta_{a_j} \zeta_{a_i} H^{diff}(\omega_j, \omega_i)\exp\left[i(\omega_j - \omega_i)t\right], \qquad (9.30)$$

where ζ_{a_j}, ζ_{a_i} are the regular wave amplitudes and H^{diff} is the difference quadratic transfer function (QTF), which is a complex number having real and imaginary part (amplitude and phase).

Numerical tools, such as WAMIT and WADAM, can be used to calculate the QTF (refer to Karimirad 2013). Such second-order problem requires the discretization of the free surface in addition to the structure-wetted surface. WAMIT V6.3S (WAMIT 2013) has the capability to account for difference-frequency components of the second-order forces and moments (QTF), the second-order hydrodynamic pressure on the body and in the fluid domain, the second-order wave elevation and the second-order response amplitude operator (RAO), all in the presence of bichromatic and bidirectional waves and one or more structures. The floating, constrained or fixed offshore structures can be considered. Considering the bichromatic and bidirectional waves is important for short crested sea states because the QTF also depends on the directions of propagation of the wave components.

The difference frequency loads (both slowly varying and mean drift) can excite the low natural frequency of the floating offshore structures, i.e. for spar and semi-submersible platforms. These structures usually have a small water plane area, which results in low natural frequency for heave motion. Also, vertical drift forces (in heave motion) may be important in shallow water areas.

In general, increasing the water surface area results in more hydro loads. Also, for multi-column platforms, the increase of distance between columns reduces the interference effects between columns and hence less hydro loads may appear. For multi-body systems, the momentum approach gives the total drift force. However, the direct pressure integration of second-order fluid pressure is required to calculate individual mean drift forces.

Newman approximation:

All pairs of frequencies (ω_j, ω_i) may contribute to the second-order difference frequency wave loads. However, the second-order forces are important when they are close to natural periods of the structure. For floating structures with slowly varying responses, the force components with difference frequencies close to the natural frequency are important. This means difference frequencies should be equal to the natural frequency. Hence, two lines in the $\omega_j \omega_i$-plane represent: $\omega_j - \omega_i = \pm\omega_N$.

For floating offshore structures with very low natural frequencies, i.e. for spar platform 0.05 rad/sec for surge/sway natural frequencies, it can be assumed that $\omega_j = \omega_i$. Newman approximation drives the off-diagonal terms of QTF using diagonal terms as follows:

$$H^{diff}(\omega_j, \omega_i) = 0.5\left[H^{diff}(\omega_j, \omega_j) + H^{diff}(\omega_i, \omega_i)\right]. \qquad (9.31)$$

In 1974, when Newman presented this simplification, the computation of QTF was demanding. Now, refined computation of full QTF is reasonable considering the computational time and book keeping of data. However, still this method is a fast approach to get slowly varying loads for structures with very low natural frequencies. The accuracy of the method is usually higher for horizontal motions due to lower frequencies. The method can be applied to TLPs, Spars and Semi-submersibles. Caution is needed when the QTF is not smooth close to the diagonal i.e. for heave motion of spar (DNV 2007).

3. Sum-frequency loads

Sum-frequency loads are second-order wave forces in an irregular sea-state oscillating at the $\omega_j + \omega_i$ frequencies. These loads can excite high-frequency resonant responses of marine structures, i.e. heave/roll/pitch of TLPs or global elastic responses of ships. The stationary time-harmonic oscillation of structural responses of marine structures is called springing. Springing is a periodic resonant excitation of structural vibration. For ships, the encounter frequency considering the forward speed is measured and linear, sum-frequency and triple-frequency springing may occur. For offshore structures, springing loads are essential for the prediction of fatigue of TLP tethers. The sum-frequency loads in an irregular sea state consisting of N regular wave with the frequency of $\omega_i \, i = 1:N$ oscillate at difference frequencies $\omega_j + \omega_i$:

$$F^{sum}(t) = \text{Re}\sum_{j=1}^{N}\sum_{i=1}^{N} \zeta_{a_j} \zeta_{a_i} H^{sum}(\omega_j, \omega_i)\exp\left[i(\omega_j + \omega_i)t\right], \qquad (9.32)$$

where ζ_{a_j}, ζ_{a_i} are the regular wave amplitudes and H^{sum} is the sum QTF, which is a complex number having real and imaginary part (amplitude and phase).

In numerical calculation of QTF, the discretization (mesh) of wetted floater geometry and free surface as well as number of frequency pairs in the QTF matrix are important factors affecting the accuracy. Convergence study, sensitivity analysis and numerical tests are required to ensure that the structure and free-surface mesh are refined. Also, to capture the second-order interaction effects between columns of multi-column structures, i.e. TLPs, a very fine frequency mesh must be used for short waves (high frequencies). Both diagonal and off-diagonal terms should be considered when selecting wave periods, as there may be peaks outside the diagonal.

9.8.7   Higher-Order Wave Loads

Higher-order wave loads can excite high-frequency resonant responses of floating offshore platforms, i.e. vertical motions of tensioned buoyant platforms (like TLPs). Also, bottom-fixed offshore structures, i.e. slender gravity-based structures (GBS), can be excited in high-frequency resonant elastic structural responses.

Vertical motions (heave, roll and pitch responses) of a TLP have high-frequency resonant responses due to stiff tendons, i.e. the heave eigenperiod is in the range 2–5 s. First-order wave loads do not excite such structures in resonant response. Note: the natural periods of floating structures are usually set out of wave-energy spectrum.

However, the structure may be excited by waves with periods $2T_N$, $3T_N$,... which carry more energy. Due to nonlinear wave effects and nonlinear fluid-structure interaction effects, there is a nonlinear transfer of energy to higher-order (super-harmonic) response of the structure. Hence, regular waves of frequency a excite the structural response at $2\omega, 3\omega, ...$

As it is mentioned before, the high-frequency stationary time-harmonic oscillation is called springing while large resonant high-frequency transient response is called 'ringing'. Ringing exciting waves have a wave length considerably longer than a characteristic cross section of the structure (Faltinsen 1999). Basic studies on ringing loads on a fixed vertical and infinitely long circular cylinder in deep water incident waves were reported by Faltinsen, Newman and Vinje (FNV), refer to (Faltinsen et al. 1994). Ringing loads are inertia loads and should not be confused by slamming loads. Ringing occurrences are more likely in extreme sea states. Time domain hydroelastic analysis considering higher-order wave loads are necessary to investigate the ringing phenomena.

Bibliography

Bak, C., Zahle, F., Bitsche, R., Kim, T., Yde, A., Henriksen, L. C., Hansen, M. H., Blasques, J., Gaunaa, M., & Natarajan, A. (2013). *The DTU 10-MW reference wind turbine*. Denmark: Technical University of Denmark, DTU Wind Energy.

Burton, T., Sharpe, D., Jenkins, N., & Bossanyi, E. (2008). *Wind energy handbook*. England: Wiley.

Chella, M. A., Tørum, A., & Myrhaug, D. (2012). An overview of wave impact forces on offshore wind turbine substructures. *Energy Procedia, 20,* 217–226.

DNV. (2007). *Environmental conditions and environmental loads*. Norway: Det Norske Veritas.

DNV. (2011). *Modelling and analysis of marine operations*. Norway: Det Norske Veritas.

DNV. (2013). *Design of offshore wind turbine structures*. Norway: Det Norske Veritas AS.

Faltinsen, O. M. (1993). *Sea loads on ships and offshore structures*. UK: Cambridge University Press.

Faltinsen, O. M. (1999). Ringing loads on a slender vertical cylinder of general cross-section. *Journal of Engineering Mathematics, 35,* 199–217.

Faltinsen, O. M., Newman, J. N., & Vinje, T. (1994). *Nonlinear wave loads on a slender vertical cylinder. Journal of Fluid Mechanics,* 289, 179 -198.

Hansen, M. O. (2008). *Aerodynamics of wind turbines* (2nd ed.). UK: Earthscan.

Jiang, Z. Y., Karimirad, M., & Moan, T. (2012). *Response analysis of a parked spar-type wind turbine under different environmental conditions and blade pitch mechanism fault*. Greece: ISOPE.

Jonkman, J. M. (2007). *Dynamics modeling and loads analysis of an offshore floating wind turbine*. USA: National Renewable Energy Laboratory (NREL).

Karimirad, M. (2011). *Stochastic dynamic response analysis of spar-type wind turbines with catenary or taut mooring systems* (PhD thesis). Norway: NTNU.

Karimirad, M. (2012). Mechanical-dynamic loads. In A. Sayigh (Ed.), *Comprehensive renewable energy* (pp. 243–268). UK: Elsevier.

Karimirad, M. (2013). Modeling aspects of a floating wind turbine for coupled wave–wind-induced dynamic analyses. *Renewable Energy, 53,* 299–305.

Karimirad, M., & Moan, T. (2010). *Effect of aerodynamic and hydrodynamic damping on dynamic response of spar type floating wind turbine*. Proceedings of the EWEC2010, European Wind Energy Conference. Warsaw: EWEA.

Karimirad, M., & Moan, T. (2012). Stochastic dynamic response analysis of a tension leg spar-type offshore wind turbine. *Journal of Wind Energy (Wiley)*. doi:10.1002/we.

Karimirad, M., Meissonnier, Q, Gao, Z, Moan, T. (2011). Hydroelastic code-to-code comparison for a tension leg spar-type floating wind turbine. *Marine Structures, 24*(4), 412–435.

Korotkin, A. I. (2009). *Added masses of ship structures*. Germany: Springer.

Lee, C. H. (1995). *WAMIT theory manual*. USA: MIT.

Manwell, J. F., McGowan J. G., & Rogers A. L. (2006). *Wind energy explained, theory, design and application*. Chichester: Wiley

Maruo, H. (1960). The drift of a body floating on waves. *Journal of Ship Research, 4,* 1–10.

Newman, J. N. (1977). *Marine hydrodynamics*. Cambridge: MIT Press.

Roshko, A. (1960). Experiments on the flow past a circular cylinder at very high Reynolds numbers. *Fluid Mechanics, 10,* 345–356.

Tangler, J. L., & Somers, D. M. (1995). NREL Airfoil Families for HAWTs. USA: AWEA.

WAMIT. (2013). *WAMIT user manual*. USA: WAMIT INC.

Chapter 10
Dynamic Response Analyses

10.1 Introduction

In the previous chapters, the loads and load cases important for offshore energy structures are discussed. To assess the functionality and structural integrity of a design, it is needed to predict the motion and structural responses. A reliable and robust design should be based on accurate calculation of loads and responses. Offshore energy structures are complicated, respect to the dependency of loads and load-effects. In these cases, the response itself may also be important for the loads, i.e. hydro-elastic effects and coupled effects between floater and mooring system. The wave- and wind-induced loads are highly connected to instantaneous wave elevation, relative motions and responses. Hence, the instantaneous position should be considered for updating the hydrodynamic and aerodynamic forces. Depending to the structure and its characteristics, the moving structure should use the accelerations and velocities at the instantaneous position. In some cases, the geometrical updating adds some nonlinear loading that can excite the natural frequencies of the structure. The relative velocity should be applied to the hydro loads and the updated wave acceleration at the instantaneous position is required for analysing some concepts. Definitely, dynamic response analysis is the base for design of offshore structures. In some cases, limit states analyses are based on combinations of individual dynamic analysis, i.e. consider a FLS which is based on accumulated damages. This shows the importance of performing correct dynamic analyses for offshore energy structures including the wave power, wind energy and hybrid energy devices.

10.2 Dynamics of Single Degree of Freedom Systems

Dynamic response of offshore structures are complicated, hence, comprehensive methods are needed to analyse the motion and structural responses of them under wave, current and wind loading. However, it is very useful to start understanding

© Springer International Publishing Switzerland 2014   223
M. Karimirad, *Offshore Energy Structures*, DOI 10.1007/978-3-319-12175-8_10

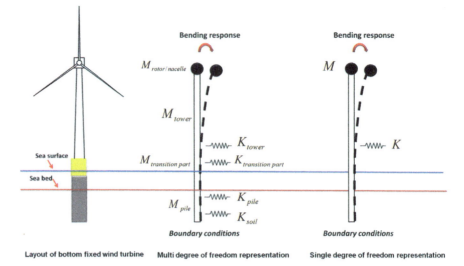

Fig. 10.1 A bottom-fixed offshore wind turbine and representation of tower-top displacement as multiple and single degree of freedom for bending response

the vibrations of single degree of freedom (SDOF) systems. In several examples, it is possible to isolate a specific response of an offshore energy structure and consider the system in that mode as a SDOF system. First, lets us look to the following examples.

Figure 10.1 illustrates a bottom-fixed offshore wind turbine and simplified representation of it using multiple and single degree of freedom presentation of bending response. The wind turbine tower-foundation can be assumed as a beam (or beams, depending to level of simplification) with elastic boundary conditions at soil–pile interface. The rotor-nacelle assembly can be considered as a point mass at tower-top or a mass matrix. Later, we see that it is possible to find a formula for land-based wind turbine eigen-period using single degree of freedom representation for structural elastic responses. Figure 10.2 illustrates a spar-type offshore wind turbine and simplified representation of heave motion response using single degree of freedom. The mooring mass and pre-tension are normally negligible compared to total mass of the system for catenary moored spar-type wind turbine. Hence, the buoyancy force can be assumed to be equal to total mass of the system for such system $(Mg = \rho g \nabla)$. In general, for catenary moored floating structures (particularly for spar platforms), the mooring stiffness is small in heave motion (the main mooring stiffness is in horizontal direction, for surge/sway motion responses). Hence, the heave stiffness is mainly coming from hydrostatic restoring forces due to structure section area at MWLS (Figs. 10.1, 10.2).

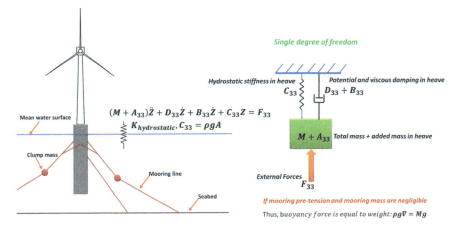

Fig. 10.2 A floating offshore wind turbine and single degree of freedom representation for heave motion response. Normally, mooring pre-tension and mooring mass is assumed to be negligible for heave motion response of such catenary moored spar platform

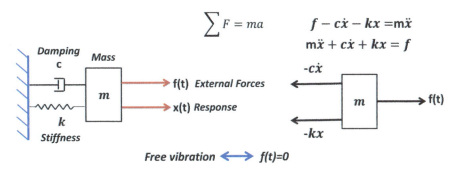

Fig. 10.3 Single degree of freedom system and its equation of motion

10.2.1 Free Vibration of Single Degree of Freedom Systems

Herein, free response of single degree of freedom (SDOF) systems is explained. A single degree of freedom system is (an idealized) one degree of freedom spring-mass-damper system. The degree of freedom can be either rotational or translational, and the mass is allowed to move in only one direction. It is assumed that the spring has no damping or mass, the mass has no stiffness or damping, and the damper has no stiffness or mass, see Fig. 10.3. For example, the horizontal vibrations of a single-storey building may be modelled as a single degree of freedom system.

Let us consider the single degree of freedom system presented in Fig. 10.3. First, we define Lagrange equations for conservative systems. The generalized forces can be derived from a potential energy function (E_p) in a conservative system. Potential energy functions present the effects of: ideal springs and gravity. For these systems, a so-called 'Lagrangian' L is defined ($L = E_K - E_P$).

The Lagrange equations for conservative systems are given by:

$$\frac{d}{dt}\left(\frac{\partial L}{\partial \dot{q}}\right) - \frac{\partial L}{\partial q} = Q \tag{10.1}$$

In which, q is noting to generalized motion and Q is indicating generalized forces. For SDOF defined in Fig. 10.3, the kinetic, potential energies and generalized forces are:

$$E_K(x, \dot{x}) = 0.5m\dot{x}^2 \tag{10.2}$$

$$E_P(x) = 0.5kx^2 \tag{10.3}$$

$$Q = f - c\dot{x}$$

Hence, Lagrange equation for such system is given by:

$$\frac{d}{dt}\frac{\partial E_K}{\partial \dot{x}} - \frac{\partial E_K}{\partial x} + \frac{\partial E_P}{\partial x} + c\dot{x} - f = 0 \tag{10.4}$$

Finally, the equation of motion for SDOF system can be represented by:

$$m\ddot{x}(t) + c\dot{x}(t) + kx(t) = f(t), \ x(0) = x_0 \quad \text{and} \quad \dot{x}(0) = v_0 \tag{10.5}$$

It is possible to derive the equation of motion by balancing the inertia, stiffness, damping and external forces and considering the Newton laws. The solution of equation of motion has two parts: a free response (homogeneous part) and a forced response (particular part).

For free vibrating system, external force is zero ($f(t) = 0$). We assume response in the form of $x(t) = Xe^{\mu t}$ and insert it in the equation of motion, by some math we get:

$$\left(m\mu^2 + c\mu + k\right)Xe^{\mu t} = 0.$$

The non-trivial solution is $m\mu^2 + c\mu + k = 0$ which has two roots:

$$\mu_{1,2} = -\frac{c}{2m} \pm \sqrt{\left(\frac{c}{2m}\right)^2 - \frac{k}{m}}$$

The initial displacement and initial velocity are used to determine the coefficients X_1 and X_2 for solution of this second order ordinary differential equation (ODE) corresponding to μ_1 and μ_2. The dynamic response, μ_1 and μ_2 are strongly affected by the amount of the system damping (c). The following cases are possible depending to damping magnitude.

1. Undamped system, $c = 0$

For undamped case when system has no damping, $\mu_{1,2} = \pm i\sqrt{\dfrac{k}{m}} = \pm i\omega_n$ in which ω_n is the natural frequency of the system. The dynamic response solution is in the form of:

$$x(t) = X_1 e^{i\omega_n t} + X_2 e^{-i\omega_n t} = A\sin\omega_n t + B\cos\omega_n t \tag{10.6}$$

2. Critically-damped system, $c = c_c$

If the system is critically damped, $c = c_c = 2\sqrt{mk}$, then $\mu_1 = \mu_2 = -\dfrac{c}{2m} = -\omega_n$. The ratio of the damping to the critical damping is called the damping ratio $(\zeta = c/c_c)$. The response solution that satisfies arbitrary initial displacements and velocities is

$$x(t) = X_1 e^{-\omega_n t} + X_2 t e^{-\omega_n t}. \tag{10.7}$$

3. Over-damped system, $c > c_c$

For over-damped case when damping is greater than the critical damping, the roots μ_1 and μ_2 are distinct and real. The over-damped system does not freely oscillate and the solution is in the form of:

$$x(t) = X_1 e^{\mu_1 t} + X_2 e^{\mu_2 t}. \tag{10.8}$$

4. Under-damped system, $0 < c < c_c$

For under damped system, the system oscillates from some initial displacement and velocity. The roots are complex conjugates and the solution is in the form of

$$x(t) = X e^{\mu t} + X^* e^{\mu^* t} \tag{10.9}$$

in which, $(*)$ denotes complex conjugate.

We may rewrite the equation of motion as: $\ddot{x}(t) + 2\zeta\omega_n \dot{x}(t) + \omega_n^2 x(t) = f(t)/m$ and the expression for the roots as: $\mu_{1,2} = -\zeta\omega_n \pm \omega_n\sqrt{\zeta^2 - 1}$.

In offshore technology, it is likely to have an under-damped system, i.e. consider the free-decay tests for moored structures in ocean basin. Herein, the response of the system given the initial value for displacement and velocity is considered in more detail. It is possible to set over-damped and critically-damped system as a special case of under-damped case.

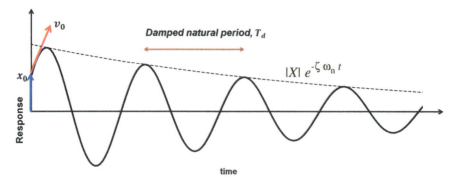

Fig. 10.4 Decaying oscillation for an under-damped system

For an undamped system, $\zeta < 1, \sqrt{\zeta^2 - 1}$ is imaginary and $\mu_{1,2} = -\zeta\omega_n$ $\pm i\omega_n\sqrt{|\zeta^2 - 1|}$. The $\omega_n\sqrt{|\zeta^2 - 1|}$ is so-called 'damped natural frequency', (ω_d). Damped natural frequency is the frequency at which under-damped SDOF systems freely oscillate. So, we rewrite the solution in the following form using the damped natural frequency, natural frequency and damping ratio.

$$x(t) = e^{-\zeta\omega_n t}\left(A\sin\omega_d t + B\cos\omega_d t\right) \tag{10.10}$$

Using the initial displacement and velocity, the A and B can be found, hence,

$$x(t) = e^{-\zeta\omega_n t}\left(\frac{v_0 + \zeta\omega_n x_0}{\omega_d}\sin\omega_d t + x_0\cos\omega_d t\right) \tag{10.11}$$

Decay test can be performed in ocean basins to find useful information about natural frequency and damping of the system. The initial displacement (i.e. surge) is defined and the platform freely oscillates, see Fig. 10.4. Usually, the initial velocity is set to zero for such tests, $v_0 = 0$

$$x(t) = e^{-\zeta\omega_n t}\left(\frac{\zeta x_0}{\sqrt{|\zeta^2 - 1|}}\sin\omega_d t + x_0\cos\omega_d t\right) \tag{10.12}$$

The logarithmic decrement method (Gavin 2014) is used to calculate the damping ratios as a function of two succeeding response amplitudes ($X_i > X_{i+1}$ for a system with positive-damping).

$$\zeta = \frac{c}{c_c} = \delta/\sqrt{\pi^2 + \delta^2}, \text{ in which, } \delta = \ln\left(\frac{X_i}{X_{i+1}}\right) \tag{10.13}$$

10.2.2   Forced Vibration of Single Degree of Freedom Systems

When structures are subjected to external loads, their responses compose of two parts: steady state and transient. The transient response decays with decay frequency (ω_d) while steady state response oscillates with external load frequency (ω). Herein, a SDOF dynamic system subjected to a harmonic forcing, $f(t) = F\cos\omega t$ applied to the mass, with forcing-frequency (ω) is considered. After several cycles, the system responds only at the external forcing-frequency, if external force is persistent. The corresponding harmonic steady-state response can be assumed as: $x(t) = A\cos\omega t + B\sin\omega t$. If we substitute the assumed function into the equation of motion, then:

$$m\omega^2(-A\cos\omega t - B\sin\omega t) + c\omega(-A\sin\omega t + B\cos\omega t)$$
$$+ k(A\cos\omega t + B\sin\omega t) = F\cos\omega t \tag{10.14}$$

This equation can be represented as two parts (sin and cos parts) and, consequently, the A and B can be found.

$$(-m\omega^2 A + c\omega B + kA)\cos\omega t = F\cos\omega t \quad \text{and}$$
$$(-m\omega^2 B - c\omega A + kB)\sin\omega t = 0 \tag{10.15}$$

$$A = \frac{k - m\omega^2}{\left(k - m\omega^2\right)^2 + (c\omega)^2}F \quad \text{and} \quad B = \frac{c\omega}{\left(k - m\omega^2\right)^2 + (c\omega)^2}F \tag{10.16}$$

It is possible to write the solution in the form of $x(t) = A\cos\omega t + B\sin\omega t = X\cos(\omega t + \varphi)$ in which the amplitude of motion (X) is $\sqrt{A^2 + B^2}$. The phase between applied force and response is: $\varphi = \dfrac{-B}{A} = -\dfrac{c\omega}{k - m\omega^2}$

The response always lags the external forcing (φ is negative). The ratio of response/force is expressed below and has unit of flexibility (m/N).

$$\left|\frac{x(t)}{f(t)}\right| = \frac{1}{\sqrt{\left(k - m\omega^2\right)^2 + (c\omega)^2}} \tag{10.17}$$

Assuming response and external force in the form of $x(t) = Xe^{i\omega t}$ and $f(t) = Fe^{i\omega t}$, it is possible to re-derive the above expressions using complex exponential notation.

$$\frac{X}{F} = \frac{1}{\left(k - m\omega^2\right) + i(c\omega)} \tag{10.18}$$

The static displacement (X_{st}) is $\dfrac{F}{k}$, hence, we may re-write the expression as following.

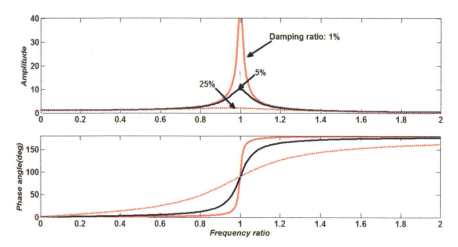

Fig. 10.5 Transfer function and phase angle versus frequency ratio for different damping ratios

$$\frac{X}{X_{st}} = \frac{1}{\left(1-\varpi^2\right)+i\left(2\zeta\varpi\right)}, \quad \varpi = \frac{\omega}{\omega_n} \tag{10.19}$$

The ratio of dynamic response amplitude and static amplitude is so-called 'dynamic amplification factor', $\mathrm{DAF} = \left|\dfrac{X}{X_{st}}\right|$. DAF is the factor by which static displacement responses are amplified for dynamic external forcing.

Figure 10.5 shows the amplitude of transfer function for motion of a SDOF system as well as the phase angle for different damping ratios. For lower damping ratios, the response is larger at the natural frequency of the system due to resonance.

10.3   Natural Periods of Floating Structures

As it is shown in the previous sections, the response amplitude operator, motion transfer function, dynamic amplification factor, and consequently, the response magnitude are increased at the natural frequency and at frequencies close to natural eigen-frequency. This highlights the importance of defining clearly the natural frequencies and eigen modes in order to set them out of load and excitation frequencies as far as it is possible.

We need to review the load frequencies to continue the discussion for natural periods. As, the natural frequencies are become important respect to load/excitation frequencies and possibility of encountering of these frequencies with a natural frequency. Due to external loading and operation, there are cyclical forces/moments applied to offshore energy structures and marine renewable energy devices. The frequency at which these forces occur can be considered as forcing frequencies with respect to the natural frequencies and harmonic responses. The loading frequencies can be fixed, random or a function of another frequency. The most significant loading-frequencies are fluid action on the support structure, waves and the cyclic passing of turbine blades or other moving components of the energy converter. Some of the excitations are broadband excitations as they comprise of frequencies, i.e. more than one frequency (EMEC 2009).

The structural and mechanical elements have a lot of natural frequencies and eigen modes. The natural frequencies which could be excited by forcing should be identified. The mode shapes of all natural frequencies must be considered to understand how such frequencies may be excited. For a single degree of freedom, it was shown that the natural frequency is a simple relation between mass and stiffness $\omega_n = \sqrt{k/m}$. However, for multi-degree of freedom systems this is complicated. Still in some cases, it is possible to isolate the desired degree of freedom and try to find the natural frequency. This may be applied to obtain the natural frequencies of rigid body motion of the floating structures.

The common practice in design of offshore structures (oil and gas) is to avoid resonance due to first hydrodynamic loading (as much as possible) by setting the natural frequencies out of the wave spectrum. This simply means that the natural frequencies are out of the common wave frequency range at the offshore site. The typical first order wave loading is between 0.2 to 1.2 rad/s. The offshore wind turbines are similar in this sense as the intention is to minimize the support structure deflections, displacement or motions. So, same practice may be applied to set the natural frequencies of the system out of the first order wave loading. However, higher order wave loads still can hit the natural frequencies and influence the response, performance and integrity. However, some of these higher order action/action-effects can be controlled/reduced using damping. Reference is made to Fig. 10.5, in which, the effect of damping ratio effect is clearly shown for a SDOF.

When it comes to wave energy converters, there are mainly two types of devices: (a) the devices that exploit a resonant response to maximize the captured-power, and, (b) those that do not (EMEC 2009). For devices that apply resonant response to maximize the power captured, the design basis should note the fatigue stresses and the fatigue life through FLS considerations. It is important to identify the natural frequencies to set them close to wave spectral peak frequencies. Example of such device is a heaving point absorber. It is possible to control and reduce the resonant responses by the following approaches:

a. Set the natural frequency of the structural and mechanical elements sufficiently separated from the forcing frequency:
   I. Either stiffness of the structural/mechanical elements is set low such that the natural frequency will occur sufficiently below the forcing frequency; or

II. Stiffness of the structural/mechanical elements is set relatively high such that the natural frequency will occur above the forcing frequency.

b. Implement sufficient damping in the design such that the fatigue stresses are not significant, i.e. in comparison to mean stresses.

Using the approach (I) results in the lowest cost as less material is needed. When the first natural frequency is below any forcing frequency, the second (or third etc.) modes should also be checked to be far from the forcing frequencies. The approach (II) is the simplest method of avoiding a harmonic response but is not a cheap option. The approach (b) is difficult to achieve unless a separate damping component is added (EMEC 2009).

Normally, surge, sway and yaw natural periods for a moored offshore structure are more than 100 s. This allows slowly varying motion of the horizontal degrees of freedom and avoiding first order wave excitation by setting the natural frequencies smaller than the cut-in wave frequencies. For spar-type floating wind turbines, the yaw natural frequency is usually set above the cut-out wave frequency (i.e. yaw natural period of 5 s) to get proper responses under wind-induced yaw excitations. Generally, heave, roll and pitch natural periods are above 20 s. For a Tension Leg Platform (TLP), and similar buoyant tethered platform and some taut moored buoys, the natural periods for vertical motions (heave, roll and pitch) are typically below 5 s. This ensures setting the natural frequencies above the cut-off wave frequencies.

As it is mentioned before, the motions of moored offshore structures are coupled. However, it is possible to apply SDOF considerations and by accounting for the hydrostatic stiffness, mooring stiffness, mass and added mass, derive the undamped natural frequencies. In most cases, the undamped natural periods derived by this approach are consistent with the experiment and numerical results.

$$\omega_j = \sqrt{\left(C_{jj} + K_{jj}\right)/\left(M_{jj} + A_{jj}\right)} \qquad (10.20)$$

ω_j is the natural frequency of the jth motion ($j = 1,2,\ldots 6$). $C_{jj}, K_{jj}, M_{jj}, A_{jj}$ are the diagonal terms in the hydrostatic stiffness, mooring stiffness, mass and added mass matrices.

The undamped uncoupled heave natural frequency of a freely floating offshore structure $(K_{33} = 0)$ or a catenary moored structure $(K_{33} \ll C_{33})$, i.e. for a moored buoy or semisubmersible is

$$\omega_3 = \sqrt{C_{33}/\left(M + A_{33}\right)} \qquad C_{33} = \rho g S_{MWL} \qquad (10.21)$$

M is the total mass, C_{33} is the heave hydrostatic restoring, S_{MWL} is the mean water level surface, K_{33} is the mooring stiffness in heave direction and A_{33} is the added mass in heave direction.

For a tension leg platform (TLP), the stiffness of tendons is much higher than the hydrostatic stiffness $\left(K_{33} \gg C_{33}\right)$, hence,

$$\omega_3 = \sqrt{K_{33}/\left(M + A_{33}\right)} \qquad K_{33} = EA/L, \qquad (10.22)$$

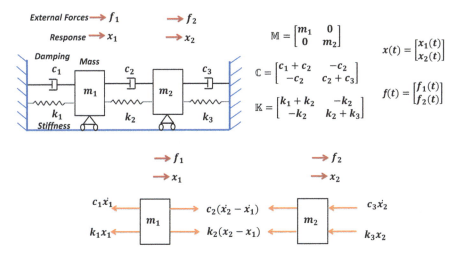

Fig. 10.6 Vibration of a two degree of freedom dynamic system

where E is the modulus of elasticity, A is the tendons total cross-sectional area and L is the length of each tendon.

The pitch natural frequency of a freely floating structure, i.e. a ship-shaped structure (not moored, $K_{55} = 0$) is

$$\omega_5 = \sqrt{C_{55}/(I_{55} + A_{55})} \qquad C_{55} = \rho g \nabla GM_L \qquad (10.23)$$

For a moored structure where K_{55} is not negligible, the pitch natural frequency is derived as:

$$\omega_5 = \sqrt{C_{55} + K_{55}/(I_{55} + A_{55})} \qquad (10.24)$$

For catenary moored structures, it is possible to relate the mooring stiffness in pitch and surge motions (as well as for roll and sway motions) using the fairlead distance to centre of gravity (KF–KG):

$$K_{55} = K_{11}(KF - KG)^2 \qquad (10.25)$$

10.4 Two Degree of Freedom System Dynamics

As an example, eigenvalue analysis of a two degree of freedom system is considered, see Fig. 10.6.

The dynamic response equation for the two degree of freedom system presented in Fig. 10.6 is as following:

$$\mathbb{M}\ddot{x} + \mathbb{C}\dot{x} + \mathbb{K}x = f \qquad (10.26)$$

$$\begin{bmatrix} m_1 & 0 \\ 0 & m_2 \end{bmatrix}\begin{bmatrix} \ddot{x}_1(t) \\ \ddot{x}_2(t) \end{bmatrix} + \begin{bmatrix} c_1+c_2 & -c_2 \\ -c_2 & c_2+c_3 \end{bmatrix}\begin{bmatrix} \dot{x}_1(t) \\ \dot{x}_2(t) \end{bmatrix}$$
$$+ \begin{bmatrix} k_1+k_2 & -k_2 \\ -k_2 & k_2+k_3 \end{bmatrix}\begin{bmatrix} x_1(t) \\ x_2(t) \end{bmatrix} = \begin{bmatrix} f_1(t) \\ f_2(t) \end{bmatrix} \tag{10.27}$$

For the free vibration analysis of the system, the external forces are set to zero. Further, if the damping is disregarded, the equations of motion reduce to:

$$\begin{bmatrix} m_1 & 0 \\ 0 & m_2 \end{bmatrix}\begin{bmatrix} \ddot{x}_1(t) \\ \ddot{x}_2(t) \end{bmatrix} + \begin{bmatrix} k_1+k_2 & -k_2 \\ -k_2 & k_2+k_3 \end{bmatrix}\begin{bmatrix} x_1(t) \\ x_2(t) \end{bmatrix} = 0 \tag{10.28}$$

We are interested to find out harmonic oscillations of masses with the same frequency and phase angle but with different amplitudes. Hence, it is possible to set the eigenvalue problem for the system as:

$$-\omega^2\begin{bmatrix} m_1 & 0 \\ 0 & m_2 \end{bmatrix}\begin{bmatrix} X_1 \\ X_2 \end{bmatrix} + \begin{bmatrix} k_1+k_2 & -k_2 \\ -k_2 & k_2+k_3 \end{bmatrix}\begin{bmatrix} X_1 \\ X_2 \end{bmatrix} = 0 \tag{10.29}$$

The eigenvalue problem represents two simultaneous homogeneous algebraic equations in the unknowns. For a nontrivial solution, the determinant of coefficients of unknowns must be zero.

$$\det\begin{bmatrix} -m_1\omega^2+k_1+k_2 & -k_2 \\ -k_2 & k_2+k_3-m_2\omega^2 \end{bmatrix} = 0 \tag{10.30}$$
$$m_1m_2\omega^4 - \left((k_1+k_2)m_2+(k_2+k_3)m_1\right)\omega^2 + (k_1+k_2)(k_2+k_3)-(k_2)^2 = 0$$

The above equation is called the frequency or characteristic equation that yields the frequencies of the characteristic values of the system, the natural frequencies (or eigen-frequencies).

$$\omega_1^2,\omega_2^2 = \frac{1}{2}\left(\frac{(k_1+k_2)m_2+(k_2+k_3)m_1}{m_1m_2}\right) \tag{10.31}$$
$$\pm\frac{1}{2}\sqrt{\left(\frac{(k_1+k_2)m_2+(k_2+k_3)m_1}{m_1m_2}\right)^2 - 4\left(\frac{(k_1+k_2)(k_2+k_3)-(k_2)^2}{m_1m_2}\right)}$$

Also, it is possible to find the response ratios $r_1 = \dfrac{X_2}{X_1}\Big|_{\omega_1}$ and $r_2 = \dfrac{X_2}{X_1}\Big|_{\omega_2}$ using the equation of motion.

$$\left(-m_1\omega^2 + (k_1 + k_2)\right)X_1 - k_2 X_2 = 0 \tag{10.32}$$

$$\left(-m_2\omega^2 + (k_2 + k_3)\right)X_2 - k_2 X_1 = 0 \tag{10.33}$$

$$r_1 = \frac{-m_1\omega_1^2 + (k_1 + k_2)}{k_2} = \frac{k_2}{-m_2\omega_1^2 + (k_2 + k_3)} \tag{10.34}$$

$$r_2 = \frac{-m_1\omega_2^2 + (k_1 + k_2)}{k_2} = \frac{k_2}{-m_2\omega_2^2 + (k_2 + k_3)} \tag{10.35}$$

The normal modes of vibration, modal vector of the system, are defined as following:

$$x^1(t) = \begin{bmatrix} x_1^1(t) \\ x_2^1(t) \end{bmatrix} = \begin{bmatrix} X_1^1\cos(\omega_1 t + \varphi_1) \\ X_2^1\cos(\omega_1 t + \varphi_1) \end{bmatrix}$$
$$= \begin{bmatrix} X_1^1\cos(\omega_1 t + \varphi_1) \\ r_1 X_1^1\cos(\omega_1 t + \varphi_1) \end{bmatrix} \equiv first\ mode \tag{10.36}$$

$$x^2(t) = \begin{bmatrix} x_1^2(t) \\ x_2^2(t) \end{bmatrix} = \begin{bmatrix} X_1^2\cos(\omega_2 t + \varphi_2) \\ X_2^2\cos(\omega_2 t + \varphi_2) \end{bmatrix}$$
$$= \begin{bmatrix} X_1^2\cos(\omega_2 t + \varphi_2) \\ r_2 X_1^2\cos(\omega_2 t + \varphi_2) \end{bmatrix} \equiv second\ mode \tag{10.37}$$

For a two degree of freedom system, the motion for each mass is a combination of first mode and second mode motions $x(t) = x^1(t) + x^2(t)$.

$$x_1(t) = X_1^1\cos(\omega_1 t + \varphi_1) + X_1^2\cos(\omega_2 t + \varphi_2) \tag{10.38}$$

$$x_2(t) = X_2^1\cos(\omega_1 t + \varphi_1) + X_2^2\cos(\omega_2 t + \varphi_2)$$
$$= r_1 X_1^1\cos(\omega_1 t + \varphi_1) + r_2 X_1^2\cos(\omega_2 t + \varphi_2) \tag{10.39}$$

X_1^1, X_1^2, φ_1 and φ_2 are found from initial conditions. Note that the initial displacement and velocity for both masses are known. Hence, four algebraic equations for four unknowns can be expressed to find the unknowns.

$$x_1(0) = x_{01} \quad x_2(0) = x_{02} \qquad (10.40)$$

$$\dot{x}_1(0) = v_{01} \quad \dot{x}_2(0) = v_{02} \qquad (10.41)$$

10.5   Eigen-Value Analysis of Multi Degree of Freedom Systems

Engineering systems are normally continuous and hence, they have infinite number of degrees of freedom. The vibration analyses of continuous systems need partial differential equations which are demanding and complicated. Actually, for many partial differential equations, analytical results do not exist. However, continuous systems are often approximated as multi degree of freedom systems. The analysis of a multi degree of freedom system requires the solution of a set of ordinary differential equations. For N-degrees of freedom system, there are N associated natural frequencies and N corresponding mode shapes.

Different approaches are introduced to simplify a continuous system as a multi degree of freedom system.

1. Lumped mass method:
   This is a simple method replacing the distributed mass/inertia of the system by lumped masses or rigid bodies. The lumped masses are connected by massless spring and damping members.
   The motions of lumped masses are defined using linear coordinates. Degrees of freedom of the system are the minimum number of coordinate essential to define the motion of the lumped masses and rigid bodies. Higher accuracy of the analysis is obtained by using more lumped masses to define the system.
2. Finite element method:
   Another method to simplify a continuous system as a multi degree of freedom system is replacing the system by a large number of elements. Using simple solution within each element, the compatibility and equilibrium principles are applied to find the original system solution.

For the first approach, lumped mass method, the equations of motion of a multi degree of freedom system can be obtained using the Newton second law of motion. A suitable coordinate system with clear positive direction to describe the positions of the point masses (and rigid bodies) should be applied. The static equilibrium configuration of the system should be studied first. Afterward, the spring, damping and external forces acting on each mass or rigid body are defined. Hence, the Newton second law is applied and the equation of motion in matrix format will be set.

The differential equations of the system are coupled and the equations cannot be solved individually one at a time (they should be solved simultaneously). The system is statically coupled if the stiffness-matrix has at least one non-zero off-diagonal term and if the mass-matrix has at least one non-zero off-diagonal term, the system is dynamically coupled. If both the stiffness and mass matrices have non-zero off-diagonal terms, the system is both statically and dynamically coupled.

The eigenvalue analysis of the equation of motion for natural frequencies and normal modes requires a reduced form of equation of motion considering zero damping and no external loading. The equation of motion for undamped free vibration in matrix format is: $\mathbb{M}\ddot{x} + \mathbb{K}x = 0$ where, $\mathbb{M}$ and $\mathbb{K}$ are mass and stiffness matrices and x is the motion (deflection or displacement) vector. As explained for SDOF, we may assume a harmonic solution: $x = \mathbb{X}sin\omega t$ where, $\mathbb{X}$ is the eigenvector or mode shape representing the shape of the system which does not change with time; only the amplitude varies. After simplification, the equation is written in the form of

$$\left(\mathbb{K} - \omega^2 \mathbb{M}\right)\mathbb{X} = 0. \tag{10.42}$$

This is the eigen-equation and it is a set of equations for the eigenvectors (the base for eigenvalue analysis). In structural engineering, the representation of mass and stiffness in the eigen-equation results in finding the natural frequencies and mode shapes. The non-trivial solution ($\mathbb{X} \neq 0$) is obtained when $det\left(\mathbb{K} - \omega^2 \mathbb{M}\right) = 0$.

The determinant is zero for a set of discrete eigenvalues and its corresponding eigenvector. Hence, we may rewrite eigen-equation in the following form. The mode having the lowest frequency is called the first mode.

$$\left(\mathbb{K} - \omega_i^2 \mathbb{M}\right)\mathbb{X}_i = 0. \quad i = 1,2,3,\ldots \tag{10.43}$$

The number of eigenvalues and eigenvectors is equal to the number of dynamic degrees of freedom. The natural frequencies and mode shapes are useful in various dynamic analyses, for example: The deflected shape of a linear elastic structure vibrating in free or forced vibration, at any given time, is a linear combination of its normal modes considering the modal displacements (ξ_i: i^{th} modal displacement).

$$x = \Sigma_i \xi_i \mathbb{X}_i \tag{10.44}$$

Amplitudes found from the eigenvalue problem are arbitrary (any amplitude will satisfy the basic frequency equation) and only the shapes are unique. As all amplitudes of a mode shape will represent a possible equilibrium between restoring and inertia forces, the amplitude for free oscillation is undetermined.

In the analysis of the two degree of freedom system presented before in this chapter, the amplitude of first mode was set to unity, and the second mode was determined relative to it. This is normalizing the mode shapes with respect to a specified reference. In many computer programs, the shapes are normalized relative to the maximum displacement value in each mode.

An eigenvector $\mathbb{X}_i$ scaled as per some normalization condition is called a normal mode ϕ. Based on normalization criterion, the scaling factor c is chosen.

$$\mathbb{X}_i = c_i \phi_i \quad i = 1, 2, \dots n \tag{10.45}$$

Normal modes satisfy the orthogonality properties with respect to the mass and stiffness matrices:

$$\phi_i^T \mathbb{M} \phi_j = 0 \quad \phi_i^T \mathbb{K} \phi_j = 0 \quad i \neq j \tag{10.46}$$

Proof of orthogonality of modes is given herein:

$$\mathbb{K} \phi_i = \omega_i^2 \mathbb{M} \phi_i \tag{10.47}$$

$$\phi_j^T \left(\mathbb{K} \phi_i \right) = \phi_j^T \left(\omega_i^2 \mathbb{M} \phi_i \right) = \omega_i^2 \phi_j^T \mathbb{M} \phi_i \tag{10.48}$$

$$\mathbb{K} \phi_j = \omega_j^2 \mathbb{M} \phi_j \tag{10.49}$$

$$\phi_i^T \mathbb{K} \phi_j = \omega_j^2 \phi_i^T \mathbb{M} \phi_j \tag{10.50}$$

$$\left(\phi_i^T \mathbb{K} \phi_j \right)^T = \left(\omega_j^2 \phi_i^T \mathbb{M} \phi_j \right)^T \rightarrow \phi_j^T \mathbb{K} \phi_i = \omega_j^2 \phi_j^T \mathbb{M} \phi_i \tag{10.51}$$

$$\omega_i^2 \phi_j^T \mathbb{M} \phi_i = \omega_j^2 \phi_j^T \mathbb{M} \phi_i \rightarrow \left(\omega_i^2 - \omega_j^2 \right) \phi_j^T \mathbb{M} \phi_i = 0 \tag{10.52}$$

$$\phi_j^T \mathbb{M} \phi_i = 0 \quad i \neq j \tag{10.53}$$

If $i = j$ the generalized mass (modal mass), and the generalized stiffness (modal stiffness) are presented as following:

$$\phi_i^T \mathbb{M} \phi_j = M_i \quad \phi_i^T \mathbb{K} \phi_j = K_i \quad i = j \tag{10.54}$$

The natural frequencies based on generalized mass and generalized stiffness appear as:

$$\omega_i = \sqrt{K_i / M_i} \quad i = 1, 2, \dots, n \tag{10.55}$$

The normalized eigenvectors (normal modes) are said to be mass-orthonormal, if the scale factors are chosen so that $\phi_i^T \mathbb{M} \phi_j = M_i = 1$ for $i = 1, 2, \dots, n$.

There are several algorithms in mathematics to solve the eigenvalue problem. The mathematical algorithms are out of the scope of this book and are not discussed further.

10.6   Rigid Body Modes

If the system has no stiffness/supports, it moves as a rigid-body which is considered as modes of dynamic motions (vibration) with zero frequency. This semi-definite system has a singular stiffness matrix and rigid-body displacements take place without any force applied. For rigid-body modes, the element must undergo without stresses developed in the system. In general case, up to six rigid body modes are possible, i.e. a helicopter has all six possible rigid-body modes, three translations and three rotations.

In offshore technology, when we are considering floating bodies, due to hydrostatic restoring forces and mooring stiffness, the smallest natural frequencies so-called 'rigid-body natural frequencies' are not necessary zero. For example, consider a ship-shaped structure, the surge, sway and yaw natural frequencies are zero if no mooring system is applied and structure is freely floating. However, the pitch, roll and heave natural frequencies are not zero because of hydrostatic restoring for these modes. For a moored structure, i.e. a tensioned leg hybrid marine platform, all the 'rigid-body natural frequencies' are non-zero; simply, as the stiffness (hydrostatic- or mooring-stiffness or both) is not zero.

10.7   Modal Dynamic Analysis for Multi Degree of Freedom System

The equations of motions for a linear multi degree of freedom (MDOF) system subjected to external loads can be written as the following form:

$$\mathbb{M}\ddot{\mathbb{x}}+\mathbb{C}\dot{\mathbb{x}}+\mathbb{K}\mathbb{x}=\mathbb{f} \tag{10.56}$$

It is useful to transform the equations of motions to modal coordinates. The response vector of a MDOF system is presented in terms of modal coordinates:

$$\mathbb{x}(t)=\sum_{j=1}^{N}\phi_{j}q_{j}(t)=\Phi\mathbb{q}(t) \tag{10.57}$$

In which, ϕ_{j} are the natural modes of the undamped system. Inserting the modal representation of the response in the motion equations, results in:

$$\sum_{j=1}^{N}\mathbb{M}\phi_{j}\ddot{q}_{j}(t)+\sum_{j=1}^{N}\mathbb{C}\phi_{j}\dot{q}_{j}(t)+\sum_{j=1}^{N}\mathbb{K}\phi_{j}q_{j}(t)=\mathbb{f} \tag{10.58}$$

Multiplying each term of the above equation by ϕ_{i}^{T}:

$$\sum_{j=1}^{N}\phi_{i}^{T}\mathbb{M}\phi_{j}\ddot{q}_{j}(t)+\sum_{j=1}^{N}\phi_{i}^{T}\mathbb{C}\phi_{j}\dot{q}_{j}(t)$$

$$+\sum_{j=1}^{N}\phi_{i}^{T}\mathbb{K}\phi_{j}q_{j}(t)=\phi_{i}^{T}\mathbb{f} \tag{10.59}$$

Due to orthogonality principles, all terms of stiffness and mass are zero except when $i = j$.

$$\phi_i^T \mathbb{M} \phi_i \ddot{q}_i(t) + \sum_{j=1}^{N} \phi_i^T \mathbb{C} \phi_j \dot{q}_j(t) + \phi_i^T \mathbb{K} \phi_i q_i(t) = \phi_i^T \mathbf{f} \qquad (10.60)$$

$$M_i \ddot{q}_i(t) + \sum_{j=1}^{N} \phi_i^T \mathbb{C} \phi_j \dot{q}_j(t) + K_i q_i(t) = F_i(t) \qquad (10.61)$$

in which, $M_i = \phi_i^T \mathbb{M} \phi_i$ is the generalized mass, $K_i = \phi_i^T \mathbb{K} \phi_i$ is the generalized stiffness and $F_i(t) = \phi_i^T \mathbf{f}$ is the generalized force. These parameters only depend to i^{th} mode. Moreover, by defining $C_{ij} = \phi_i^T \mathbb{C} \phi_j$, we may rewrite the equation as following:

$$M_i \ddot{q}_i(t) + \sum_{j=1}^{N} C_{ij} \dot{q}_j(t) + K_i q_i(t) = F_i(t) \qquad (10.62)$$

For damped systems, the modal equations may be coupled. However, by reasonable simplifications (for many structures), the equations become uncoupled. Hence, for classical damping, the modal equations are uncoupled.

$$C_{ij} = 0 \quad i \neq j \quad \text{and} \quad C_i = 2\zeta_i M_i \omega_i$$

In which, ζ_i is the damping ratio of the ith mode.

$$M_i \ddot{q}_i(t) + C_i \dot{q}_i(t) + K_i q_i(t) = F_i(t) \qquad (10.63)$$

$$\ddot{q}_i(t) + 2\zeta_i \omega_i \dot{q}_i(t) + \omega_i^2 q_i(t) = F_i(t)/M_i \qquad (10.64)$$

Typical value of damping ratio for mechanical systems which are lightly damped is around 0.02. For offshore bottom-fixed structures such as jackets the damping ratio of 0.05 may be applied. Wind turbine rotor aerodynamic-damping is around 0.1.

The modal analysis is incorporated in numerical tools to study the dynamics of structures. For example, FAST (Jonkman and Buhl 2005) employs a combined modal and multi-body dynamics formulation to analyse the wind turbine dynamic. Flexibility in the blades and tower are characterized using a linear modal representation that assumes small deflections. FAST models flexible elements, such as the tower and blades, using a linear modal representation. The reliability of this representation depends on the generation of accurate mode shapes which are input into FAST. Finite element software such as ABAQUS and ANSYS may be used to generate the mode shapes.

10.8 Wave-Induced Responses Applying Frequency Domain Analysis

If the viscous effects are negligible compared to load and load effects obtained by potential theory, which means the fluid is assumed irrotational and inviscid, frequency domain analysis is simple to be applied to find the responses of the structure under wave loads. Large volume structures are inertia-dominated and global loads due to wave diffraction are significantly larger than the drag induced global loads. However, Morison viscous drag load for the slender members/braces in addition to the radiation/diffraction should be applied, i.e. for braces and truss elements appearing in the design.

So, for a large volume floating structure (i.e. a semisubmersible), a linear analysis may be applied for calculating the global wave frequency loads and responses. First order wave loading without implementing viscous effect is implemented. The term linear means that the loads are proportional to the wave amplitude. Hence, the responses in irregular sea state can be found using superposition method.

As it is explained in the previous chapters, Panel methods for the wetted area of the floater up to the mean water line can be applied to account for the first order wave loading. Such diffraction/radiation wave analysis provides first order excitation forces, hydrostatics, potential wave damping, added mass, first order motions in rigid body degrees of freedom and the mean drift forces/moments. The drift forces/moments are second order forces; however, they can be calculated using first order results.

Equations of motions for a moored floating structure can be written as following:

$$\left(\mathbb{M} + \mathbb{A}(\omega)\right)\ddot{x}(t) + \mathbb{B}(\omega)\dot{x}(t) + \mathbb{C}x(t) = f(t) \tag{10.65}$$

This equation, rigid body equation of motion, contains six coupled equations for three translations (surge, sway and heave) and three rotations (roll, pitch and yaw). The mass $\left(\mathbb{M}\right)$, added mass $\left(\mathbb{A}(\omega)\right)$, potential damping $\left(\mathbb{B}(\omega)\right)$, hydrostatic restoring plus mooring stiffness $\left(\mathbb{C}\right)$ are incorporated in this equation. Also, external forces and moments can be added to this equation if needed.

The solution in the form of sinusoidal harmonic response is inserted in this equation, similar to what is presented for a SDOF system. By some math, the equations are rewritten as:

$$\left[-\omega^2\left(\mathbb{M} + \mathbb{A}(\omega)\right) + i\,\omega\mathbb{B}(\omega) + \mathbb{C}\right]\mathbb{X} = \mathbb{F} \tag{10.66}$$

Hence, the complex amplitude of structure motions is obtained from the solution of 6 by 6 linear system.

$$\sum_{j=1}^{6}\left[-\omega^2\left(M_{kj} + A_{kj}(\omega)\right) + i\omega B_{kj}(\omega) + C_{kj}\right]X_j = F_k \tag{10.67}$$

It is possible to define the linear structural operator $(\mathbb{L})$ characterizing the equation of motion as:

$$\mathbb{X} = \left[-\omega^2\left(\mathbb{M} + \mathbb{A}(\omega)\right) + i\omega\mathbb{B}(\omega) + \mathbb{C}\right]^{-1} \mathbb{F} = \mathbb{L}^{-1}\mathbb{F} \qquad (10.68)$$

$$\mathbb{L}(\omega) = \mathbb{F}\mathbb{X}^{-1} = \left[-\omega^2\left(\mathbb{M} + \mathbb{A}(\omega)\right) + i\omega\mathbb{B}(\omega) + \mathbb{C}\right] \qquad (10.69)$$

Normally, the linear wave analysis is performed for unit wave amplitude. Linear transfer functions (LTF) are used in frequency domain analysis to present different variables involved, i.e. exciting forces/moments and motions per unit wave amplitude. First order wave forces/moments are described in the frequency domain as a transfer function between wave elevation and force/moment:

$$\mathbb{F}(\omega) = \mathbb{H}^{(1)}(\omega)\varsigma(\omega) \qquad (10.70)$$

$\mathbb{H}^{(1)}(\omega)$: Complex first order force transfer function
$\quad \varsigma(\omega)$: Complex harmonic wave

The linear motion transfer function or response amplitude operator (RAO) gives the response per unit amplitude of wave:

$$RAO = X(\omega)/\varsigma(\omega) \qquad (10.71)$$

$$RAO(\omega) = \frac{X}{\varsigma} = \frac{\mathbb{L}^{-1}\mathbb{F}}{\varsigma} = \mathbb{L}^{-1}(\omega)\mathbb{H}^{(1)}(\omega) \qquad (10.72)$$

This means by knowing the linear structural operator and force/moment transfer function, the response amplitude operator (response transfer function) and consequently motion can be defined. The motions in irregular wave are a superposition of responses in regular waves. The stochastic analyses are explained in the next chapter. However, it is needed to recall some stochastic theoretical backgrounds for spectral analysis (Newland 2005).

The autocorrelation function can be defined by:

$$r_{xx}(\tau) = \lim_{T\to\infty} \frac{1}{T}\int_0^T x(t)x(t+T)dt \qquad (10.73)$$

The power spectral density function is the Fourier transform of the autocorrelation function:

$$S_{xx}(\omega) = Fourier\left(r_{xx}(\tau)\right) = \frac{1}{2\pi}\int_{-\infty}^{+\infty} r_{xx}(\tau)\exp(-i\,\omega\tau)d\tau \qquad (10.74)$$

$$r_{xx}(\tau) = InversFourier\left(S_{xx}(\omega)\right) = \int_{-\infty}^{+\infty} S_{xx}(\omega)\exp(i\,\omega\,\tau)\,d\omega \qquad (10.75)$$

Note that the wave spectrum is a one-sided spectrum which is defined as:

$$S_{xx}^{+}(\omega) = \begin{cases} S_{xx}(\omega) & \omega > 0 \\ 0 & \omega < 0 \end{cases} \qquad (10.76)$$

$$X(\omega) = RAO(\omega)\varsigma(\omega) \qquad (10.77)$$

$$S_{xx}(\omega) = \frac{1}{2\pi}\int_{-\infty}^{+\infty} \lim_{T\to\infty} \frac{1}{T}\int_{0}^{T} x(t)x(t+T)dT\exp(-i\omega\tau)d\tau \qquad (10.78)$$

$$x(t) = X(\omega)\exp(-i\omega t) \qquad \text{and} \qquad x(t+T) = X(\omega)\exp\left(-i\omega(t+T)\right) \qquad (10.79)$$

$$x(t)x(t+\mathcal{T}) = (RAO(\omega))^2\,\varsigma(\omega)\,\exp(-i\omega t)\,\varsigma(\omega)\,\exp\left(-i\omega(t+\mathcal{T})\right)$$

$$= (RAO(\omega))^2\,\varsigma(t)\,\varsigma(t+\mathcal{T}) \qquad (10.80)$$

$$S_{xx}(\omega) = \frac{1}{2\pi}\int_{-\infty}^{+\infty} \lim_{T\to\infty} \frac{1}{T}\int_{0}^{T} (RAO(\omega))^2\,\varsigma(t)\varsigma(t+\mathcal{T})\,d\mathcal{T}\,\exp(-i\omega\tau)\,d\tau$$

$$= (RAO(\omega))^2\,S_{\varsigma\varsigma}(\omega) \to S_{Res}(\omega) = (RAO(\omega))^2\,S_{Wave}(\omega) \qquad (10.81)$$

The above finding is very useful and helps to define the response characteristics in a given sea-state by solving the equation of motion for regular waves and superposition afterward. Short-term response statistics can be estimated using the response spectrum if the equations of motion and the excitation are linear. The variance of a variable is autocorrelation function for zero delay $\mathcal{T} = 0$ which is the area under its spectrum (MARINTEK 2012, SIMO theory manual):

$$\sigma_x^2 = r_{xx}(0) = \int_{-\infty}^{+\infty} S_{xx}(\omega)d\omega \qquad (10.82)$$

$$\sigma_{Res}^2 = \int_{0}^{+\infty} (RAO(\omega))^2\,S_{Wave}(\omega)d\omega \qquad (10.83)$$

There are nonlinear hydrodynamic effects presenting for marine structures like drag loads, damping and excitation, time varying geometry, restoring forces and variable surface elevation. However, these nonlinearities can be linearized in several cases (DNV 2007).

When linearization gives satisfactory results, i.e. for moderate environmental conditions, frequency domain analysis is useful to calculate motions and forces of floating structures. It is very practical for fatigue analyses due to fast and straight forward computations relative to time domain and hybrid analyses methods. However, for offshore energy structures having wind turbine included or for extreme load and response calculation of structures (as the nonlinearities increase) the time domain methods are essential and demanding. Another example is the horizontal stiffness nonlinearities. As the mooring stiffness is a function of offset of the structure, the coupling effects between platform motions and mooring system are important. Due to drift motions and slowly varying motions induced by wind and wave loads, the stiffness of mooring lines are changing and hence, time domain solutions are needed to cover such effects; the issue is more critical for shallow and moderate water depths in which the mooring lines are generally become more stiff and mooring line force-displacement relations are more nonlinear (Karimirad and Moan 2012c).

For several structures, the drag forces on the slender elements are important and should be added. Viscous drag forces are normally presenting as linear and quadratic terms. The quadratic viscous drag forces should be linearized when frequency domain analyses are carried out. In general, a nonlinear term $\left|\dot{u}\right|^{n}\dot{u}$ can be linearized in the form of $B\dot{u}$ where

$$B = \sqrt{\frac{2}{\pi}} 2^{\frac{n+1}{2}} \, \Gamma\left(\frac{n+3}{2}\right) \sigma_{\dot{u}}^{n} \tag{10.84}$$

In which, $\sigma_{\dot{u}}$ is the root mean square of the $\dot{u}$. For quadratic drag force, $n=1$, and the linearization become: $\sqrt{\frac{8}{\pi}}\sigma_{\dot{u}}\dot{u}$. Note that proper iterations are needed as $\sigma_{\dot{u}}$ is a function of responses. In other words, the loads and responses are tightly linked, for more information refer to literatures.

10.9 Response Equations for Offshore Energy Structures

Classical mechanics perfectly present the dynamic behaviour of an offshore energy structure. The response equations are represented by Newton's second law: $\mathbb{M}\ddot{\mathbb{x}} = \sum \mathbb{f}(t,\mathbb{x},\dot{\mathbb{x}})$ in which the generalized force vector includes all the loads, i.e. environmental forces including wave and wind loads, gravitational forces including gyroscopic forces, mooring system and soil interaction, stiffness and damping forces including the aerodynamic, hydrodynamic and structural stiffness and damping. $\mathbb{M}$ is the mass matrix, and $\mathbb{x}$ is the position vector including translations and rotations.

We start with a floating marine structure; the rigid body equations of motions in regular waves can be written as:

$$M\ddot{x} + B\dot{x} + D_1\dot{x} + D_2\dot{x}\left|\dot{x}\right| + Cx = f(t, x, \dot{x})$$
$$M = m + A(\omega),$$
$$A(\omega) = A_\infty + a(\omega), \quad A_\infty = A(\omega = \infty)$$
$$B(\omega) = B_\infty + b(\omega), \quad B_\infty = B(\omega = \infty) = 0 \tag{10.85}$$

Where, M is the frequency-dependent mass matrix, m is the structure mass matrix, A is the frequency-dependent added mass matrix, B is the frequency-dependent potential damping matrix, D_1 is the linear viscous hydrodynamic damping matrix, D_2 is the quadratic viscous hydrodynamic damping matrix, C is the position-dependent hydrostatic stiffness matrix, x is the structure position vector including translations and rotations, and f is the force vector.

The force vector (f) includes wind forces (excitation and damping), wave excitation forces including first order, second order (mean drift, slowly varying, sum and difference frequency) and higher order hydrodynamic loads, current drag forces (damping presents through hydrodynamic-damping in the left-hand side), wave drift damping, mooring system forces, specified external forces, coupling effect loads as well as the gravitational and buoyancy forces (for a free floating structure at equilibrium, $(mg = \rho Vg)$. Other forces, such as earthquake forces, ice loading and etc., can be added depending to the offshore site and the concept.

For an example, if we consider a free floating buoy in heave motion, in calm water and neglecting the damping forces, one may write:

$$M\ddot{z} + Cz = -mg + \rho Vg \tag{10.86}$$

For a deep slender buoy, the heave added mass may be neglected compared to the total mass. Also, the heave restoring forces are related to surface area (S). For an increased draft (d), the equation is rewritten as:

$$m\ddot{z} + \rho gSz = \rho gSd \tag{10.87}$$

Remembering the solution for a SDOF system:

$$z(t) = e^{-\zeta\omega_n t}\left(\frac{v_0 + \zeta\omega_n z_0}{\omega_d}\sin\omega_d t + z_0\cos\omega_d t\right) \Rightarrow z(t) = -d\cos\sqrt{\frac{\rho gS}{m}}t$$
$$\zeta = 0, \quad z_0 = -d, \quad v_0 = 0 \tag{10.88}$$

10.9.1   Floating Wind Turbines Aero-Loads Considerations

For floating wind turbines, the aerodynamic drag forces on the tower considering the relative wind velocity account for both the excitation and damping aero-loads. Hence, the main differences compared to offshore oil/gas platforms are the wind loads on the rotor, gyroscopic effects and rotating rotor existence, controller actions and power take-off through the generator. Herein, these points are briefly discussed.

Case A: Parked Shutdown Turbine If the turbine is parked, the loads are very similar to those for a regular floating marine structure. The rotor is subjected to drag aerodynamic loads and by accounting for the relative velocity, the aerodynamic damping of the parked rotor is also considered. The mass matrix should accounts for the tower and rotor mass/inertia contributions. The rotor is stand-still, hence, no gyroscopic and centrifugal loads present. In this case, wind turbine is similar to a topside-structure on the platform which is subjected to wind loads, refer to (Karimirad and Moan 2010).

Case B: Idle Shutdown Turbine If the wind turbine is idle (having rotor rotating, but, disconnected from the generator), the gyroscopic and centrifugal forces should be included. The lift and drag forces appear on the blades which results in both aerodynamic excitations and damping forces by accounting for the relative velocities. The mass matrix should consider mass/inertia of the rotating rotor. As the rotor is rotating, its mass distribution is similar to a disc and this modelling should be accurate enough for inclusion in the mass matrix of rigid-body equations of motions. If finite element modelling is used, distributed mass at each time step considering the accurate position of the elements are implemented. However, the intension can be to simplify the problem and use the rigid-body modelling with limited degrees of freedom (i.e. six dofs).

Case C: Operating Turbine If the wind turbine is operating, the modelling and forces are similar to those for Case B, except that the controller effects and power-take off should be accounted for. In the above-rated wind speed, for a floating wind turbine, negative damping may also present. Considerations are needed if conventional PI-controller is applied for a floating wind turbine. In the simplified modelling using rigid body equations of motions, the filtering of unwanted frequency components is needed to get rid of negative damping, refer to (Karimirad and Moan 2012). For a comprehensive modelling with a controller, the control parameters are tuned to avoid such negative damping effects (Karimirad and Moan 2011).

10.9.2 Simple Vs. Comprehensive Aero-Loads Modelling

To perform a coupled wave-wind-induced analysis for floating wind turbines, simultaneous wave and wind loads should be applied. The comprehensive modelling tools usually use blade element momentum (BEM) theory for blades and tower elements accounting for nonlinear-advanced aerodynamics, i.e. turbulence wind, shear effects, dynamic inflow and aero-elastic effects. The loads are calculated at each time step and applied on the structural elements while wave loads are updated and applied to the relevant wet elements. However, there are limited numerical tools custom-made for such purposes. Sometimes the users start from hydrodynamic codes and try to add important aerodynamic features.

The simplest approach is to consider a constant thrust force at the top of tower. Thrust force at each environmental condition can be calculated from a BEM code. The time-varying thrust may be applied to consider more aerodynamic points like turbulence, dynamic flow and wind shear effects. In these simple models, the wave-induced aerodynamics is missing.

The intermediate level modelling is to calculate an integrated wind loads and apply it at the top of tower. The drag forces at the tower are separately considered as an integrated wind loads at the tower. The advantage here is to account for the platform wave-wind-induced motions when updating the wind loads at each time steps; this is done by applying relative velocities. The negative damping should be removed for above-rated wind speed regions which can be performed by filtering approach. The thrust coefficients used in this method are pre-calculated by a BEM code. This method may be useful when dealing with complicated hybrid marine energy platforms having both wind and wave energy devices.

10.9.3 Wave Energy Converters Considerations

The power take-off system of a wave energy converter is usually modelled as a spring-damper system. The parameters of the power take-off unit should be pre-calculated by another numerical tool or using experiments. In some cases, the wave energy converter can be modelled as single body with six degree of freedoms, i.e. consider a heaving buoy point absorber. However, there are wave energy devices that should be modelled using multi-body hydrodynamic approach considering the kinematic relations between the bodies and hydrodynamic interaction between the structures. An example can be a semisubmersible platform with heaving buoys or flaps as wave energy devices. Another example is the spar platform having a torus around it at the mean water surface which takes power from waves by heaving up and down. In such case, the system can be modelled with two rigid bodies linked by power-take-off (PTO) system. Hydrodynamic interaction between bodies can be important depending to size of bodies and distance between them. Multi degree of freedom (MDOF) systems have been explained in this chapter and a two degree of freedom system has been represented in detail as an example. Due to importance of multi-body dynamics for hybrid marine platforms and wave energy converter, the hydrodynamic aspects of such systems will be explained in more details.

10.9.4 Solution Methods for Rigid-Body Response-Equations

Convolution Integral

The frequency-dependent coefficients included in the response-equations make it challenging to solve the dynamic response equations. One of the most-used methods for solving is based on convolution integrals. Considering the radiation part of the problem and applying the inverse Fourier transform, the radiation part can be related to the retardation function, $R(t)$, which is calculated using either potential damping or added mass; for instance, refer to (Falnes 2005).

$$(m + A_\infty)\ddot{x} + D_1\dot{x} + D_2\dot{x}|\dot{x}| + Cx + \int_0^t R(t-\tau)\dot{x}(\tau)d\tau = f(t, x, \dot{x}) \qquad (10.89)$$

$$R(t) = \frac{2}{\pi}\int_0^\infty b(\omega)\cos\omega t\, d\omega = -\frac{2}{\pi}\int_0^\infty \omega a(\omega)\sin\omega t\, d\omega$$

$$a(\omega) = A(\omega) - A(\omega = \infty)$$

$$b(\omega) = B(\omega) - B(\omega = \infty) \equiv B(\omega)$$

$$(10.90)$$

The rewritten equations of motion using retardation functions are known as the 'Cummins' equations which contain the memory effect of the generated waves using retardation formulation. This method has been applied in many engineering codes and numerical tools dealing with coupled response analysis of floating marine structures.

For large volume structures like semisubmersibles and ship-shaped structures, i.e. FPSOs, added mass and potential damping are highly frequency-dependent. However, for slender marine structures like Spar platforms, the potential damping and added mass is almost frequency-independent which means the retardation function, $R(t)$, converges to zero. The rigid-body equations of motion for slender marine structures can be written without the convolution integral which is consistent with the Morison formula.

Separation of Motions

SIMO (MARINTEK 2012) has an alternative solution for solving the response-equations in time domain by separating motions/responses in high-frequency and low-frequency parts. The high-frequency part of response is solved in the frequency domain assuming that the motions are linear responses to waves. The exciting force is separated to wave-frequency (f^{WF}) and low-frequency (f^{LF}) parts. The low-frequency (f^{LF}) excitations include wind drag force, current drag force, second-order wave forces and other forces.

$$f(t, x, \dot{x}) = f^{WF}(t, x, \dot{x}) + f^{LF}(t, x, \dot{x}) \qquad (10.91)$$

Also, the position vector is separated to high-frequency (x^{WF}) and low-frequency (x^{LF}) parts: $x(t) = x^{WF}(t) + x^{LF}(t)$

The wave-frequency motions in frequency-domain are expressed as:

$$(m + A(\omega))\ddot{x}_{WF} + (D_1 + B(\omega))\dot{x}_{WF} + Cx_{WF} = f^{WF}(\omega) \qquad (10.92)$$

The wave-induced responses applying frequency domain analysis, using the transfer function of first-order wave forces, have been presented earlier in this chapter. The low-frequency responses contain nonlinear terms such as quadratic drag forces; hence, the equations of motions for this part should be solved in time-domain.

$$\left(m + A(\omega = 0)\right)\ddot{x}_{LF} + D_1\dot{x}_{LF} + D_2\dot{x}_{LF}\left|\dot{x}_{LF}\right| + Cx_{LF} = f^{LF}(\omega) \qquad (10.93)$$

10.10   Comprehensive Analysis of Offshore Energy Structures

For an offshore energy structure, especially for floating wind turbine and hybrid marine energy platforms, nonlinear stochastic time-domain integrated coupled analysis tools that can be used for hydro-elastic-aero-servo simulations are required. As the hydrodynamic and aerodynamic loads vary over time-space and due to strong link between load and responses, the equations of motions should be solved at each time step. Moreover, the loads should be updated considering the instantaneous position and relative velocities of the structure at each time step.

The time-domain/frequency-domain, uncoupled/integrated analysis, linear/nonlinear modelling, rigid/elastic body modelling, steady/turbulent wind simulation and linear/nonlinear wave theory are options for a dynamic response analysis. Depending to the structure, functionality and response under study, a proper modelling and analysis should be applied.

Several numerical tools are available for dynamic response analysis of marine structures and renewable energy structures; among those SIMA, SIMO-RIFLEX, Bladed, HAWC2, FAST, Flex5 and USFOS codes are widely applied for offshore energy structures including wave and wind energy devices.

Different methods such as frequency-domain, time-domain and hybrid-time-frequency-domain techniques are used for dynamic analysis of marine structures. As explained, the frequency domain analysis is very fast; on the other hand, it is not always possible to use the frequency domain methods for offshore energy structures due to nonlinear wave and wind loading, control, strong coupling of rotor-platform, geometrical updating, large deformation, coupling of wave and wind loads, highly linked force-displacement relations, coupling of mooring stiffness and motions, and similar issues. In these cases, the integrated time domain analysis is necessary for such structures.

From a time domain analysis several information such as maximum, high and low frequency responses, strange peaks and very slow variations are obtained. Moreover, the time series are transformed to the frequency domain and presented in spectral format to make it easier to follow the nature of the response. The time-domain simulations should be long enough to ensure the statistical reliability. Note that the first part of the time-domain simulation is influenced by transient responses and hence, it should be eliminated before transforming to the frequency domain.

For nonlinear systems, the time domain analysis should be applied for solving the equations of motion, i.e. for a floating wind turbine. As the nonlinearities in the loading are considerable, the linearization of the equations of motion is not accurate to represent the dynamic structural responses. Even if linear elastic theory is used to model the structure, the loading and consequently, the responses are nonlinear (Karimirad 2011).

Wind loads are inherently nonlinear; the aerodynamic lift and drag-type forces are fully nonlinear. The hydrodynamic quadratic drag forces are similar to aerodynamic forces in nature and add to nonlinearities involved. The geometrical updating, force-displacement relations and nonlinear coupling of mooring system forces with motions of the structure necessitate the fully coupled time-domain response analysis of offshore energy structures in general. The mooring lines are nonlinear elastic elements; the nonlinear force-displacement or FE modelling can be used. Still, for some concepts and in special cases frequency-domain analysis can be used to provide acceptable approximate solutions.

The aerodynamic and hydrodynamic damping, wave-induced aerodynamic damping and wind-induced hydrodynamic damping should be considered for offshore energy structures. The control algorithm controls the output power. The coupled time-domain analysis is a reliable approach to account for all these issues.

The equations of motion for a floating wind turbine are nonlinear and can be solved in time domain using direct step-by-step integration techniques. Time domain analysis allows the handling of nonlinearities involved in hydrodynamic and aerodynamic loading and finite wave amplitude effects as well as nonlinear material and geometrical effects (Karimirad 2011). To summarize this discussion, some of the important points for coupled time-domain analysis of offshore energy structures are listed below:

- Nonlinear hydrodynamic loads
  - Inertial and drag forces accounting for position updating
  - Retardation and memory effects
  - Hydro-elasticity and fluid-structure interactions
  - Current loads
  - Vortex induced vibrations
  - Vortex induced motions
  - Shallow water effects and nonlinear wave kinematics
- Soil-foundation interactions
- Wind and aerodynamic forces
  - Lift and drag excitations considering the relative velocity
  - Aero-elasticity
- Damping
  - Aerodynamic damping
  - Hydrodynamic damping
  - Wave-induced aerodynamic damping
  - Wind-induced hydrodynamic damping
  - Structural damping
  - Soil damping

- Mooring system
- Structural considerations
  - Large elastic deflections
  - Rigid body movement
  - Nonlinear finite elements
- Control and servo loads

10.10.1   Elastic-Body Response-Equations

The rigid-body equations of motions are introduced before. However, in reality, the structures are elastic and for some structures, especially for slender structures, the elasticity can affect the responses. Hence, the structure is usually divided into several bodies for an elastic structure, i.e. a wind turbine. Multi-body methods considering elastic bodies connected by stiffener/damper may be used. In these methods, the elastic formulation is applied for each body while the rigid-body connections are used to link the bodies by constraints. This methodology is used in the aero-hydro-servo-elastic codes such as HAWC2 for dynamic analysis of wind turbines. Herein, static and dynamic finite element modelling (FEM) and practical solution methods for these analyses are briefly discussed.

10.10.2   Static Finite Element Analysis

When displacements and forces are not time-dependent (inertia and damping forces are zero), the dynamic equations reduce to static equations. In static finite element analysis, the static equilibrium configuration is established as the solution of the following equations:

$$F^S(r) = F^E(r) \tag{10.94}$$

where, F^S is the internal structural reaction force vector found by assembly of element contributions. The contact forces are also treated as internal reaction forces. F^E is external force vector accounting for specified external forces, contribution from distributed loading, i.e. weight, buoyancy and current forces and rigid body forces including representation of buoys, clump weights, etc. r is the structural nodal displacement vector including all degrees of freedom for the system, for example, three degrees of freedom (displacements) at each node for modelling of bar; and, six degrees of freedom (displacements and rotations) for a beam model. Displacements and rotations should refer to the stress free reference configuration. In general, the internal reaction forces and external loading are nonlinear functions of the nodal displacement vector.

The purpose of the static analysis is to determine the nodal displacement vector so that the complete system is in static equilibrium. The static equilibrium needs to be satisfied before performing the dynamic analysis. The state of the discretized finite element model is completely determined by the nodal displacement vector. Numerically, the static equilibrium is found by application of an incremental loading procedure with equilibrium iteration at each load step. A so-called incremental-iterative procedure with Euler–Cauchy incrementation method can be applied (MARINTEK 2014).

10.10.3 Dynamic Finite Element Analysis

For dynamic analysis, a full elastic representation of structures utilizing finite element methods may be used and elastic formulations implementing time incremental approaches are applied. In the following, the finite element formulation for elastic equations of motions is discussed. The dynamic equilibrium of a spatial discretized finite element model can be expressed as:

$$F^I(r,\ddot{r},t) + F^D(r,\dot{r},t) + F^S(r,t) = F^E(r,\dot{r},t), \qquad (10.95)$$

where, F^I is the inertia force vector, F^D is the damping force vector, F^S is the internal structural reaction force vector, F^E is external force vector, and $r,\dot{r},\ddot{r}$ are the structural displacement, velocity and acceleration vectors. Due to (a) the displacement dependencies of the inertia and the damping forces, (b) the coupling between external load vector and structural displacement and velocity, the equation is a nonlinear system of differential equations. In addition, there is a nonlinear relationship between internal forces and displacements. The force vectors are established by assembly of element contributions and specified discrete nodal forces (MARINTEK 2014).

The external force vector (F^E) accounts for the weight and buoyancy, wave loads, mooring system forces, forced displacements (if applicable), specified discrete nodal forces and aerodynamic loads.

The wave excitation loads accounts for drag, diffraction and Froude–Krylov forces. If Morison formula is used, the wave loads account for drag and wave acceleration terms. The aerodynamic loads including the drag and lift forces are calculated considering the instantaneous position of the element and the relative wind velocity. The blade element momentum (BEM) theory is used to present the aerodynamic loads on the tower, nacelle and rotor including the blades and hub. The aerodynamic damping forces can be kept on the right-hand side or moved to the damping force vector on the left-hand side.

The inertia force vector (F^I) can be defined by the following expression:

$$F^I(r,\ddot{r},t) = \left[M^S + M^H(r) \right]\ddot{r} \qquad (10.96)$$

where, M^S is the structural mass matrix, and $M^H(r)$ is the displacement-dependent hydrodynamic mass matrix accounting for the added mass contributions in local directions.

The damping force vector F^D is expressed as:

$$F^D(r,\dot{r},t) = \left[D^S(r) + D^H(r) + D^D(r) \right] \dot{r}, \qquad (10.97)$$

where, $D^S(r)$ is the internal structural damping matrix, $D^H(r)$ is the hydrodynamic damping matrix accounting for the radiation effects for floating and partly submerged elements, and $D^D(r)$ is the matrix of specified discrete dashpot dampers, which may be displacement-dependent.

As discussed before, by linearization of system matrices and hydrodynamic loading at static equilibrium position, frequency-domain analysis can be applied. The frequency domain analysis gives a Gaussian response described by the mean value and the response spectrum linked to the standard deviation. But, the time domain approach allows for description of Gaussian as well as non-Gaussian responses. The dynamic equilibrium equations can be solved in the time domain through step-by-step numerical integration based on Newmark-β methods or similar numerical approaches.

10.11   Multi-Body Dynamics Considering Hydrodynamic Interactions

Offshore energy structures may consist of multiple floaters. An important aspect for these structures is the response of interconnected or multiple body ocean energy structures such as wave energy converters. Several concepts of wave energy converters work based on the relative displacements between floating/submerged bodies. Hydrodynamic interactions between floaters as well as between a floater and a fixed structure influence the dynamic response of hybrid marine platforms. The hydrodynamic interactions should be analysed using radiation/diffraction methods through the multi-body approaches in hydrodynamics. WAMIT (Lee 1995) applied decomposition of the radiation-potential into components, corresponding to the modes of the rigid body motion, to account for multi-body interaction. This is done by defining the velocity potential corresponding to a particular mode of one body while the other bodies are kept stationary. Separate geometry files are needed for dealing with multiple body problems. Note that each body is considered rigid in such analysis. The N floaters are considered in an integrated system with maximum $N \times 6$ DOFs. As the multiple bodies interact mechanically, the total number of DOFs in dynamic analysis may be less than $N \times 6$ (WAMIT 2013).

For instance, consider the STC concept (Made et al. 2013a, b) in which a buoy heaves along a spar; the surge, sway, roll, pitch and yaw motions of these two bodies are same; the bodies move together for these motions. Hence, the full-system has 7 degrees of freedom in dynamic analysis. The hydrodynamic properties of the rigid bodies considering their interactions (and possibly second-order wave loads) are calculated in the frequency domain and subsequently applied in the coupled motion (wave- and wind-induced responses)-mooring analysis of the system in the time domain through retardation functions. Moreover, the heave motions of the torus and spar are related by mechanical properties of the power take off system. The WEC PTO is derived from the relative heave motion between spar and torus which can be modelled as linear spring-damper system.

For a multiple-body system, a strong hydrodynamic interaction between the floating or/and fixed structures may appear. This interaction phenomenon can become more important due to unwanted relative motion responses between bodies. An important nonlinear interaction effect is waves between the bodies that can affect the loads and motions.

If boundary element methods (Panel methods) are applied, the discretization of the wetted surfaces in the area between the structures should be fine to capture the variations in the wave. In the coupled motions, extra resonance peaks may appear through mechanical and hydrodynamic interactions. Hydrodynamic interaction should be included if the excitation loads on each structure is influenced by the presence of the other structures. Also, sheltering effect which leads to smaller motions on the lee side than on the weather side may appear (DNV 2007).

A few methods and codes can handle 3D radiation-diffraction analysis of multibody marine structures considering joint interactions. A practical approach is to calculate the hydrodynamic properties and loads for each floating body oscillating in each of the 6 DOF. Then, the radiation and diffraction forces together with other loads like the power-take-off (PTO) force, mooring system and wind loads are post-processed. Afterward, the kinematic constraints are applied to reduce the DOF to the original DOF of the oscillating device. Using this approach and employing a code similar to WAMIT, no symmetry can be applied for either the radiation or for the diffraction problem. Consequently, considerable simulations are needed when the number of bodies increases. Moreover, exporting the field particle information (i.e. pressure and wave elevation) becomes quite cumbersome as the information passes through the post-processing stage for each DOF and each body. For linear hydrodynamic load assessment of multiple floating bodies, the 'generalized modes' approach can be used to transform the problems to a single body with several DOF. This method is well-known in the response analysis of flexible bodies in waves (i.e. hydro-elasticity). The 'generalized modes' approach is used for flexible floating structures as well as multi-body hydrodynamics; the method is well explained in (Newman 1994). In this approach, all coefficients are generalized; the matrices and vectors have a dimension of $(6+L)\times(6+L)$ and $(6+L)\times1$, respectively. L is the number of flexible modes defined in addition to the six rigid body modes. For multiple bodies, all the geometries are then merged and considered as one. This 'single body' has 'pseudo-flexibility' in the sense that different parts of the body (sub-bodies) have different DOF depending on the problem. In this way, $N\times6$ mandatory DOF for the initial multi-body problem (N bodies) reduces to one single body with the original M DOF where $M<N\times6$ (Taghipour October 2008).

10.12   Some Aspects of Dynamic Response

Response of offshore structures may consist of three types of responses: quasi-static, resonant and inertia dominated responses. If the frequency of the loads is much less than the natural frequencies, the response is quasi-static. In this case, the dynamic responses are slowly varying. If the excitation frequencies are close to the natural frequencies of the system, the resonant responses may occur. Note that the resonant responses, such as rigid-body natural-frequencies, are usually avoided by adjusting the natural periods away from the wind and wave spectral peaks. However, higher-order wave loads and aerodynamic forces can excite the natural frequencies and create the resonant responses. If the loading frequencies are higher than the natural frequencies, the inertia-dominated response appears. For example, the wave-induced rigid-body motions of semisubmersible platforms are inertia dominated as the wave frequencies are greater than the floater natural frequencies.

Managing the wave-wind-induced responses helps to increase the power output and decrease the fluctuations of produced power. For wind turbines, the idea is to reduce the rigid-body motions as much as possible. For wave energy converters, depending to the type of the device, it may be helpful to increase the body motions, i.e. for heaving-buoy point-absorber.

The wind loads excite the low-frequency natural periods of the floating devices. For floating energy structures, the aerodynamic damping due to operation of the wind turbine can reduce the resonant responses. The aerodynamic damping is not significant for a parked or idling wind turbine as the turbine does not generate power. Note that power generation can be simplified and presented as a damper. Controlling the blade feathering implicitly controls the floater motions. Less floater motions help to produce more electrical power. Also, the controller in variable-speed wind turbines is tuned to skip possible rotor-rotational-frequencies around the rigid body and elastic natural frequencies. This helps to reduce fatigue damage to a large extent.

It is possible to decrease the wave-induced responses by reducing the projected area of the support structures against the waves in splash zone where the hydrodynamic loads are maximal. Some support structures have advantage of having the main part below the splash zone. The other option to decrease the motions is to increase the inertia of the structure. However, this option is a costly approach due to added steel used and construction works.

The hydrodynamic damping may be increased to reduce the resonant responses by damping plates, vortex-suppression strakes or buoyancy cans at the water line. It is necessary to note the possibility of increasing the excitation forces when applying these damping features. Also, some of these features change the added mass and consequently change the natural periods of the system. Proper dynamic response analysis and experiments are needed to investigate the effects (Karimirad 2011).

For some energy structures, the flexible mode response is comparable to the wave frequency and rigid body resonant responses. Hence, an elastic modelling is necessary to properly present the structural responses.

The tension response and platform motions are tightly linked; several complex combinations of characteristic frequencies, first-order and higher-order wave frequencies and natural frequencies may appear in tension responses of mooring lines especially for taut mooring system and tension leg platforms. These frequencies are mixing through the nonlinear relationship of the tension leg stiffness.

The fatigue and ultimate limit states are two important factors in the design of ocean structures. The environmental conditions can be harsh and induce extreme responses for a floating structure. For a land-based wind turbine, fatigue is the key parameter in design, and the extreme responses that occur in operational conditions are connected to the rated wind speed. However, for a floating wind turbine, the extreme responses can occur in survival conditions.

10.13 Aero-Hydro-Elasticity Applied to Energy Platforms

Offshore energy structures (OES), Renewable ocean energy (ROE) systems, Marine renewable energy (MRE) devices, Marine Hydro Kinetic (MHK) systems, Wave and wind energy devices, Wave energy converters (WEC), Offshore wind turbines (OWT), Hybrid marine platforms (HMP) and similar names refer to devices/systems that produce electricity using wave and wind energy available in offshore sites and oceans. These devices convert the fluid potential energy, fluid kinetic energy or both to mechanical/electrical energies using aero-hydro-elastic dynamic systems. In some cases control is a key issue and, hence, the term aero-hydro-servo-elastic is commonly used in literatures.

Wind and wave loads result in elastic structural responses. As the system simultaneously responds to wave and wind actions, the aerodynamic and hydrodynamic interactions need to be accounted for in coupled simulations. An important issue for numerical-modelling and analysis of offshore energy structures is the complexity of wind and wave actions and load effects. For example, the aerodynamic loads are highly nonlinear and result from static and dynamic relative wind flow, dynamic stall, skew inflow, shear effects on the induction and effects from large deflections.

The comprehensive aerodynamic methods are based on solving the Navier–Stokes (NS) equations for the global compressible flow in addition to accounting for the flow near the blades (Sanderse 2009). The extended blade element momentum theory is normally used to consider advanced and unsteady aerodynamic effects for aero-elastic time-domain calculation for offshore energy structures. Approaches of intermediate complexity, such as the vortex and panel methods, are also applied.

The advanced blade element momentum (BEM) theory is fast and gives good accuracy compared to computational fluid dynamic (CFD) methods. To calculate the aerodynamics, the CFD methods are the most accurate (Sezer-Uzol and Long 2006) but are very time consuming. Verification and validation are needed to investigate the accuracy of numerical methods. Comparison with wind tunnel and full-scale tests are necessary to document the accuracy of these methods. The BEM method relies on airfoil data; therefore, the result obtained using this method is dependent to accuracy of airfoil data used. It is practical to use the NS methods to extract airfoil data and apply them in less advanced methods, e.g. BEM codes (Hansen et al. 2006).

The aerodynamic loads and time-dependent structural behaviour of the system are strongly coupled. A blade may change its twist and thus its angle of attack due to elastic deflections (Hansen 2008, Aerodynamics of Wind Turbines). Also, the angles of attack are changed when the blades have a velocity relative to the fixed ground. For offshore energy structures, the structural behaviour is wave-wind-induced. For instance, if the tower of a floating wind turbine is moving upstream due to wave-induced surge motion, it will be seen by the blades as an increased relative wind speed; thus, higher angles of attack will present along the blades (Karimirad and Moan 2012).

The coupled wave-wind-induced responses present aerodynamic and hydrodynamic damping influencing the dynamic responses. The motions directly affect the nacelle displacement and its velocity which influence the relative velocity and consequently the power productions. For hybrid marine platforms, the wave energy converters can reduce the support structure motions through power take off. This helps increasing the power production of the wind turbine.

The motion response frequencies, both resonant- and wave-frequency responses, appear in the relative wind velocity. The appearance of rigid-body motion's frequencies in the relative velocity influences the power production and its quality. Also, the structural responses are sensitive to the elastic body formulation and it is necessary to model the elastic body to capture accurate structural responses. In particular, large blade deflections have a significant influence on power production. Finite element modelling (FEM) with multi-body formulation is applied to model the offshore energy structures especially for wind turbines (e.g. HAWC2 code).

The modal analysis can be used to describe the structural responses by a finite series of structural eigen-frequencies (mode shapes); the accuracy of this method is highly dependent on the modes chosen to describe the structure (e.g. FAST code). If this approach is applied for floating wind turbines, the mode shapes of the tower are affected by the platform through the elasticity and added mass contributions which should be properly accounted for.

In general, offshore energy structures comprise of slender components such as tower and blades. The geometrical nonlinearities in their physical behaviour present large deflections and should be accounted for by subdividing the structure into several linear structural elements or by applying nonlinear elastic methods. The beam modelling of the slender structures can be sufficiently accurate compared to shell modelling.

The slender elements in offshore energy structures especially for offshore wind turbines suffer from low structural damping which may be critical in some conditions. Offshore wind turbines are large in size compared to land-based wind turbines; hence, investigation of their aero-elastic stability is essential. Possible aero-elastic instabilities of floating wind turbines and hybrid marine platforms should be investigated.

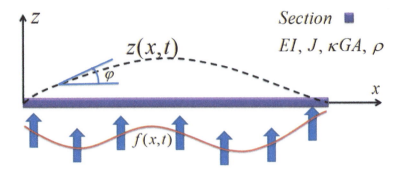

Fig. 10.7 Simple beam under dynamic load

Often in aeroelastic codes, the structure is divided to beams. The standard beam theory is linear and known as the Euler–Bernoulli beam theory. This neglects the shear effects. The relation between transversal load ($f(x,t)$) and displacement (z) can be written as:

$$\rho A \frac{\partial^2 z}{\partial t^2} + \frac{\partial^2}{\partial x^2}\left(EI \frac{\partial^2 z}{\partial x^2} \right) = f(x,t) \tag{10.98}$$

See Fig. 10.7, in which, x is the direction on the beam, ρ is the beam mass density, A is the sectional area of the beam, E is the modulus of elasticity and I is the area moment of inertia. Timoshenko beam theory accounts for shear deformation and rotational inertia effects. The dynamic equations using Timoshenko beam theory are (Bauchau and Craig 2009):

$$\rho A \frac{\partial^2 z}{\partial t^2} = \frac{\partial}{\partial x}\left[\kappa AG\left(\frac{\partial z}{\partial x} - \varphi \right) \right] + f(x,t)$$

$$\rho I \frac{\partial^2 \varphi}{\partial t^2} = \frac{\partial}{\partial x}\left(EI \frac{\partial \varphi}{\partial x} \right) + \kappa AG\left(\frac{\partial z}{\partial x} - \varphi \right) \tag{10.99}$$

φ is the slope of the beam due to bending, G is the shear modulus of the beam and κ is the Timoshenko shear coefficient which varies with the geometry of the beam. If the shear modulus of the beam is very large and rotational inertia effects are neglected, the Timoshenko beam theory converges towards Euler–Bernoulli beam theory.

Timoshenko

$$\left. \begin{array}{l} \rho A \dfrac{\partial^2 z}{\partial t^2} = \dfrac{\partial}{\partial x}\left[\kappa AG\left(\dfrac{\partial z}{\partial x} - \varphi \right) \right] + f(x,t) \\[3mm] \rho I \dfrac{\partial^2 \varphi}{\partial t^2} = \dfrac{\partial}{\partial x}\left(EI \dfrac{\partial \varphi}{\partial x} \right) + \kappa AG\left(\dfrac{\partial z}{\partial x} - \varphi \right) \end{array} \right\}$$

$$\Rightarrow \rho A \frac{\partial^2 z}{\partial t^2} = \frac{\partial}{\partial x}\left[\rho I \frac{\partial^2 \varphi}{\partial t^2} - \frac{\partial}{\partial x}\left(EI \frac{\partial \varphi}{\partial x} \right) \right] + f(x,t)$$

$$if \; \frac{\partial}{\partial x}\left[\rho I \frac{\partial^2 \varphi}{\partial t^2}\right] neglected \Rightarrow \rho A \frac{\partial^2 z}{\partial t^2} = \frac{\partial}{\partial x}\left[-\frac{\partial}{\partial x}\left(EI \frac{\partial \varphi}{\partial x}\right)\right] + f(x,t)$$

$$if\left(\varphi = \frac{\partial z}{\partial x}\right) \Rightarrow Euler - Bernouli \qquad (10.100)$$

As we discussed, the controller actions have a clear and distinguished role on the loads, responses and consequently on the performance of energy devices. A proportional-integral-derivative (PID) controller is commonly used to control off-shore energy structures. The PID controller is a simple and very practical which is extensively used in marine energy applications. This is a generic control loop feedback mechanism attempting to minimize the error. The difference between the measured and desired value of a process-variable is minimized by adjusting the process control inputs. The three constants (proportional, integral and derivative constants) are tuned in the PID controller algorithm to provide servo-actions for a specific process according to the nature of the system.

For wind turbines, the PI-controller has been widely applied and works very well. The control algorithm limits the generated power to keep the drivetrain torque constant for wind speed higher than the rated wind speed. Above a certain cut-off value (i.e. 25 m/sec), control algorithm shuts down the wind turbine. To control a wind turbine, the following basic methods are used: pitching the blades, controlling the generator torque, applying the high speed shaft (HSS) brake, deploying the tip brakes and yawing the nacelle (Jonkman and Buhl 2005). Aero-hydro-elastic codes such as Bladed, HAWC2 and FAST utilize their control modules applying dynamic link library (DLL). Effect of different control algorithms on the dynamic response can be easily tested through DLLs.

Changing the controller target from constant-power to constant-torque can reduce the aerodynamic loads on the structure (Larsen and Hanson 2007). Negative damping may appear for floating wind turbines if improper controller is applied. Larsen and Hanson showed a method for avoiding negatively damped low-frequency motion of a floating pitch-controlled wind turbine (spar buoy concept) . Negative damping in a floating wind turbine may be introduced by the blade pitch control of an operating turbine. When the relative wind speed experienced by the blades increases due to the rigid body motion of the system, the blades are feathered to maintain the rated electrical power for over-rated wind speed cases. The remedy to remove the servo-induced instabilities is modifying the controller gains for over-rated wind speed cases (Karimirad and Moan 2011).

10.14   Flutter: An Aeroelastic Dynamic Behaviour

To show an example of aero-elasticity in dynamic response analysis of offshore energy structures, flutter is briefly discussed here. Flutter is a complex dynamic phenomenon that involves the coupling of two (or more) degrees of freedom;

normally, the flapwise and the torsion degrees of freedom. Flutter is an aeroelastic instability dynamic behaviour. The torsional mode and flapwise mode are being coupled through the aerodynamic forces; and, the flutter mode appears. The aerodynamic loads lead to torsion of the blade. The torsion changes the angle of attack and the aerodynamic lift force. If the changed lift forces (due to torsion) has particular phase compared to the flapwise bending, flutter occurs. Flutter appears as violent responses with fast growing amplitude. The flutter mode has a negative damping and the structural damping is not sufficient to compensate it. If the relative wind speed is higher than a certain value (the critical flutter speed), the flutter happens. The critical flutter speed limit is the wind speed at which the aeroelastic system oscillates harmonically without further excitation after an initial disturbance. It is important to investigate the possibility of flutter for wind turbines; in particular, for larger wind turbines due to the more flexibility of the blade and the increased relative wind velocity at the blade tip, flutter may become a dimensioning criterion; refer to (Hansen 2007, Aeroelastic instability problems for wind turbines), (Vatne 2011), (Strømmen 2005).

Let us consider a simple case for a 2D blade section shown in Fig. 10.8. The elastic axis is located with a length $c.a_{CG}$ in front of the centre of gravity. c is the chord length and a_{CG} is a length factor. The aerodynamic centre is located with a length $c.a_{AC}$ in front of centre of elasticity (torsional point); refer to (Hansen 2007, Aeroelastic instability problems for wind turbines) for more information. The equations of motion for a 2D blade section shown in Fig. 10.8, neglecting the structural damping, are described as:

$$m\ddot{z} - mca_{CG}\ddot{\theta} + k_{fw}z = f_L$$
$$-mca_{CG}\ddot{z} + mc^2(r_{CG}^2 + a_{CG}^2)\ddot{\theta} + k_t\theta = ca_{AC}f_L \qquad (10.101)$$

r_{CG} is the radius of gyration about centre of gravity normalized with chord length. $I_{CG} = mc^2 r_{CG}^2$ is the mass moment of inertia.

In this example, the edgewise degree of freedom is not considered. A model of a blade section with both spring and damper using all three degree of freedom can be presented. The aerodynamic damping is the main source of damping for wind turbines. However, the structural damping of composite blades should be accounted for in the aeroelastic models.

The quasi-steady aerodynamic lift per unit length is $f_L = 0.5\rho c V_{rel}^2 C_L(\alpha)$, in which, the relative velocity can be expressed as: $V_{rel} = \sqrt{V^2 + \dot{z}^2}$. Hansen presented the angle of attack and lift force assuming that the inflow is presumed parallel to the chord as (Hansen 2007, Aeroelastic Instability Problems for Wind Turbines):

$$\alpha = \arctan\left[\frac{V\sin\theta - \dot{z} - c(0.5 - a_{AC})\dot{\theta}}{V\cos\theta}\right] \qquad (10.102)$$

The linear approximation of the lift force about $\theta = \dot{h} = \dot{\theta} = 0$:

$$f_L = f_{L_0} + \frac{1}{2}c\rho V^2 C_L' \left[\theta - \frac{\dot{z}}{V} - (0.5 - a_{AC})\frac{c\dot{\theta}}{V}\right] \qquad (10.103)$$

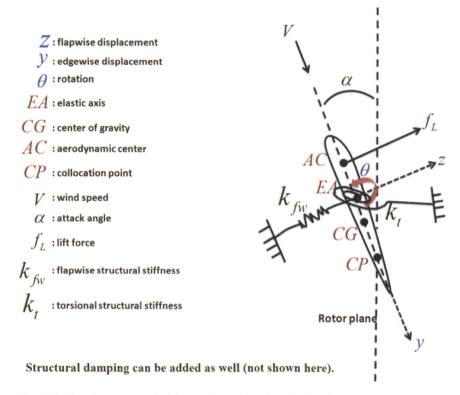

Z : flapwise displacement

y : edgewise displacement

θ : rotation

EA : elastic axis

CG : center of gravity

AC : aerodynamic center

CP : collocation point

V : wind speed

α : attack angle

f_L : lift force

k_{fw} : flapwise structural stiffness

k_t : torsional structural stiffness

Structural damping can be added as well (not shown here).

Fig. 10.8 Coordinate system for blade section with springs in flapwise and torsional direction. Structural damping is not shown

C_L and C_L' are evaluated at zero angle of attack; for thin airfoils $C_L' = 2\pi$. Also, the steady state lift (f_{L_0}) has no effect on the stability and by neglecting the camber of the airfoil: $f_{L_0} = 0$. The equations of motions for the 2D airfoil section in the matrix format for the response $\mathbf{x} = \begin{bmatrix} \dfrac{z}{c}, \theta \end{bmatrix}^{\mathrm{T}}$ and assuming the linear lift may be presented as: $\mathbb{M}\ddot{\mathbf{x}} + \mathbb{C}\dot{\mathbf{x}} + \mathbb{K}\mathbf{x} = 0$ where, the structural mass matrix, aerodynamic damping matrix and aeroelastic stiffness matrix are presented by the following expressions.

$$M = \begin{bmatrix} 1 & -a_{CG} \\ -a_{CG} & r_{CG}^2 + a_{CG}^2 \end{bmatrix} \quad C = \frac{c\,\kappa}{V} \begin{bmatrix} 1 & 0.5 - a_{AC} \\ a_{AC} & a_{AC}(0.5 - a_{AC}) \end{bmatrix}$$

$$K = \begin{bmatrix} \omega_{fw}^2 & -\kappa \\ 0 & r_{CG}^2 \omega_t^2 - \kappa a_{AC} \end{bmatrix} \tag{10.104}$$

The flapwise and torsional modes natural frequencies are: $\omega_{fw} = \sqrt{\dfrac{k_{fw}}{m}}$ and $\omega_t = \sqrt{\dfrac{k_t}{mc^2 r_{CG}^2}}$, in which, $\kappa = \dfrac{\rho}{2m} V^2 C_L'$ is aerodynamic stiffness.

For high relative inflow speeds (V) and moderate frequencies of section vibrations, the aerodynamic damping is an order smaller compared to stiffness. To study the flutter mechanism and by neglecting the aerodynamic damping, an eigen-value problem can be set which results in the following characteristic equation:

$$r_{CG}^2 \lambda^4 + \left[\left(r_{CG}^2 + a_{CG}^2 \right) \omega_{fw}^2 + r_{CG}^2 \omega_t^2 - \kappa(a_{AC} + a_{CG}) \right] \lambda^2$$
$$+ \, \omega_{fw}^2 \left(r_{CG}^2 \omega_t^2 - \kappa a_{AC} \right) = 0 \tag{10.105}$$

The zeroes of the characteristic equation are the eigen-values which generally are complex. If the real part of the eigen-value is positive, then the instability appears due to the fact that the response grows exponentially. The real part of all zeroes of the polynomial is negative if the coefficients are positive:

$$\left[\left(r_{CG}^2 + a_{CG}^2 \right) \omega_{fw}^2 + r_{CG}^2 \omega_t^2 - \kappa(a_{AC} + a_{CG}) \right] > 0$$
$$\text{and } \omega_{fw}^2 \left(r_{CG}^2 \omega_t^2 - \kappa a_{AC} \right) > 0 \tag{10.106}$$

The flutter happens if $\left(r_{CG}^2 + a_{CG}^2 \right) \omega_{fw}^2 + r_{CG}^2 \omega_t^2 \le \kappa(a_{AC} + a_{CG})$, hence, the flutter limit of the airfoil 2D section is:

$$\frac{\rho}{2m} V^2 C_L' < \frac{\left(r_{CG}^2 + a_{CG}^2 \right) \omega_{fw}^2 + r_{CG}^2 \omega_t^2}{\left(a_{AC} + a_{CG} \right)} \quad for \quad a_{AC} + a_{CG} \ge 0; \tag{10.107}$$

The inequality must be turned for $a_{AC} + a_{CG} < 0$. This simple analytical expression of the flutter limit derived for a typical section confirms the risk of flutter for wind turbines. The second criterion defines the divergence limit of the airfoil $\left(r_{CG}^2 \omega_t^2 - \kappa a_{AC} > 0 \right)$: $0.5c \, \rho V^2 C_L' ca_{AC} < k_t$. Beyond this limit, an increase in torsion will increase the lift which again increases the torsion, leading to divergence (Hansen 2007, Aeroelastic instability problems for wind turbines).

10.15   Case Study: Analysis of a Jacket Wind Turbine

Here, a jacket wind turbine subjected to wave and wind loads are studied. The model of the system in USFOS software (USFOS 2014) is presented in Fig. 10.9. The wind loads are applied as integrated thrust force and tower drag. The rotor aerodynamic loads are summed up in one point and set at the top of tower. The aerodynamic forces on tower are applied as drag force at the centre of tower projected area. The rotor/nacelle is defined as mass matrix at the top of tower. The water depth is 40 m. Wave loads are calculated using Morison formula for slender elements. It is assumed that the wind and wave loads can be uncoupled for the jacket wind turbine (Gao et al. 2010). However, this is not correct for floating wind turbine and a coupled time domain analysis is needed (Karimirad and Moan 2012).

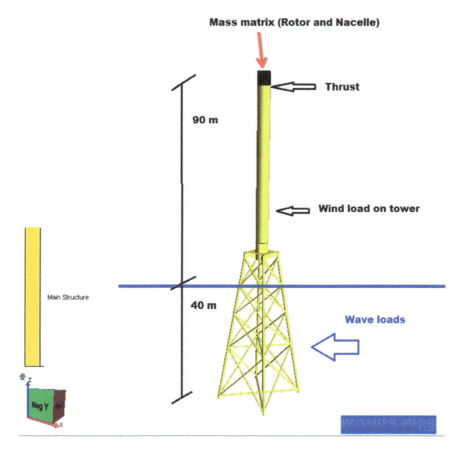

Fig. 10.9 Model of a jacket wind turbine in USFOS, the rotor/nacelle is presented as mass matrix at top of the tower

A. Wave and Wind Conditions for a North Sea Site Wave and wind conditions are correlated because waves are usually wind-generated. This correlation has been considered in the study. The Statfjord site has been chosen as a representative site for a floating wind turbine park. Statfjord is an oil and gas field in the Norwegian sector of the North Sea operated by Statoil. The site location is 59.7N and 4.0E with 70 km distance from the shore. Simultaneous wind and wave measurements covering the years 1973–1999 from the Northern North Sea are used as a database. Raw data have been smoothed and fitted to analytical functions (Karimirad and Moan 2012).

B. Contour Line for a Joint Wave and Wind Distribution Figure 10.10 shows the contour line of the joint distribution for wind and waves (1-h mean wind speed and 3-h significant wave height and wave period). The wind speed is at 10 m height from mean water level.

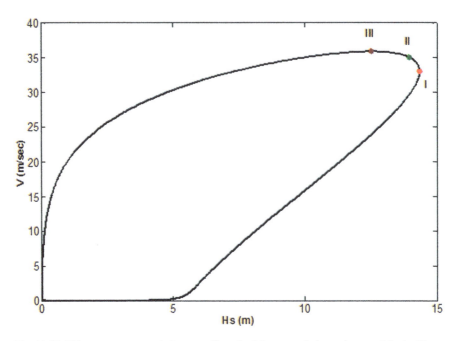

Fig. 10.10 Fifty-years return period contour lines for 1-h mean wind speed versus 3-h significant wave height

Table 10.1 Load cases; environmental conditions: wave and wind characteristics

Case number	Description	V (m/s)	Hs (m)	Tp (s)	I
L1	Operational, rated wind speed	11.4	3.0	10.0	0.15
L2	50 year environmental condition (I)	44.9	14.3	15.4	0.10
L3	50 year environmental condition (II)	47.6	13.9	15.0	0.10
L4	50 year environmental condition (III)	48.9	12.5	14.0	0.10
L5	50 year wind and corresponding sea state	51.7	13.2	14.2	0.10

C. Load Cases Based on the IEC standard and Fig. 10.10, considering the joint distributions of wave and wind, the following load cases are selected; see Table 10.1.

D. Aerodynamic Loads In this study, the wind loads are pre-calculated and used in USFOS code. The HAWC2 (DTU 2014) is used to calculate the aerodynamic loads for the load cases mentioned in Table 10.1. The aerodynamic loads/load-effects are listed in Table 10.2. The top-shear is larger than thrust force as the inertia loads are implicit in shear forces while thrust force is a pure aerodynamic load.

E. Results: Design Load Analysis The bottom shear forces and bending moments are shown in Fig. 10.11. Same as floating wind turbines, the maximum responses of jacket wind turbine occurs in survival environmental conditions. However, for a land based wind turbine the maximum responses are associated with rated wind

Table 10.2 Aerodynamic loads and load effects calculated in HAWC2, Base shear and top shear are the shear forces at the top of tower and at the bottom of tower due to wind loads. The difference between top shear and bottom shear is the drag force on the tower

Load cases	Loads (kN)	Thrust	Base shear	Top shear	Wind loads on tower
L1	Operational, rated wind speed	791	979	866	113
L2	50 year environmental condition (I)	102	885	378	507
L3	50 year environmental condition (II)	116	906	374	532
L4	50 year environmental condition (III)	123	1102	470	632
L5	50 year wind and corresponding sea state	138	1196	561	635

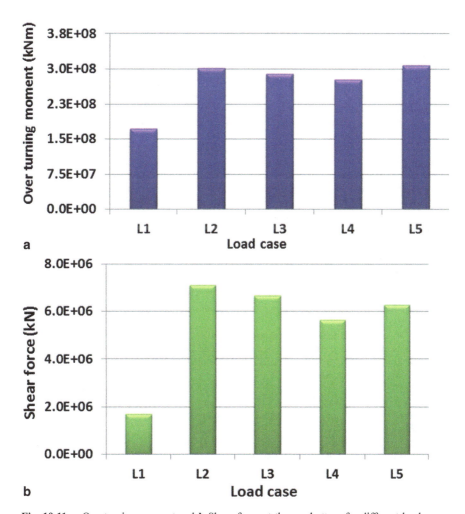

Fig. 10.11 a Overturning moment and **b** Shear force at the sea-bottom for different load cases. USFOS has been used to run dynamic analysis for the jacket wind turbine subjected to stochastic wave loads and integrated wind loads obtained from HAWC2

speed in operational conditions. It is good to mention that the FLS is the governing limit sate for jacket wind turbine rather than ULS while for a floating wind turbine the ULS can be the governing limit state depending to the concept and site.

References

Bauchau, O. A., & Craig J. I. (2009). *Structural analysis*. USA: Springer.

DNV. (2007). *Environmental conditions and environmental loads*. Norway: Recommended Practice, DNV-RP0-C205.

DTU. (2014). *DTU wind*. http://www.vindenergi.dtu.dk/english. Accessed April 2014.

EMEC. (2009). *Guidelines for design basis of marine energy conversion systems*. UK: BSI. (389 Chiswick High Road, London W4 4AL).

Falnes, J. (2005). *Ocean waves and oscillating systems*. UK: Cambridge University Press.

Gavin, H. P. (2014). *Vibrations of single degree of freedom systems*. USA: Department of Civil and Environmental Engineering, Duke University

Hansen, M. H. (2007). Aeroelastic Instability Problems for Wind Turbines. *Wind Energy, 10,* 551–577.

Hansen, M. O. (2008). *Aerodynamics of wind turbines* (2nd ed.). UK: Earthscan.

Hansen, M. O. L., Sorensen, J. N., Voutsinas, S., Sorensen, N., & Madsen, H. Aa. (2006). State of the art in wind turbine aerodynamics and aeroelasticity. *Progress in Aerospace Sciences Journal, 42,* 285–330.

Jonkman J. M., & Buhl M. L. Jr. (2005). FAST user's guide. USA: NREL. (EL-500-38230).

Karimirad, M. (2011). *Stochastic dynamic response analysis of spar-type wind turbines with catenary or taut mooring systems*. Norway: NTNU. (PHD thesis).

Karimirad, M., & Moan, T. (2010). Effect of aerodynamic and hydrodynamic damping on dynamic response of spar type floating wind turbine. *Proceedings of the EWEC2010, European Wind Energy Conference*. Warsaw: EWEA.

Karimirad, M., & Moan, T. (2011). Ameliorating the negative damping in the dynamic responses of a tension leg spar-type support structure with a downwind turbine. *Scientific proceedings of the European Wind Energy Conference (EWEC 2011)*. Brussels: EWEA.http://proceedings. ewea.org/annual2011/allfiles2/579_EWEA2011presentation.pdf. Accessed Aug 2013.

Karimirad, M., & Moan, T. (2012c). Feasibility of the application of a spar-type wind turbine at a moderate water depth. *Energy Procedia, 24*(2012), 340–350. *(DeepWind conference, 19–20 January 2012, Trondheim, Norway, published by Journal of Energy Procedia, Elsevier, Energy Procedia)*.

Larsen, T. J., & Hanson, T. D. (2007). A method to avoid negative damped low frequent tower vibrations for a floating, pitch controlled wind turbine. *Journal of Physics*. Science of Making Torque from Wind. (J. Phys.: Conf. Ser. 75 012073 doi:10.1088/1742-6596/75/1/012073) (IOP Publishing, Conference Series 75).

Lee, C.-H. (1995). *Wamit theory manual*. USA: MIT.

Made, J. M., Karimirad, M., Moan, T. (2013a). Dynamic response and power performance of a combined spar-type floating wind turbine and coaxial floating wave energy converter. *Renewable Energy, 50,* 47–57.

Made, J. M., Karimirad, M., Gao, Z., & Moan, T. (2013b). Extreme responses of a combined spartype floating wind turbine and floating wave energy converter (STC) system with survivalmodes. *Ocean Engineering, 65,* 71–82.

MARINTEK. (2012). *SIMO theory manual*. Norway: MARINTEK.

MARINTEK. (2014). *RIFLEX theory manual*. Norway: Norwegian Marine Technology Research Institute.

Newland, D. E. (2005). *An introduction to random vibrations, spectral & wavelet analysis*. USA: Dover Civil and Mechanical Engineering.

Newman, J. (1994). Wave effects on deformable bodies. *Applied Ocean Research, 16,* 47–59.

Sanderse, B. (2009). *Aerodynamics of wind turbine wakes.* Netherlands: ECN. (ECN-E-09-016).

Sezer-Uzol N., & Long L. N. (2006). 3-D time-accurate CFD simulations of wind turbine rotor flow fields. *American Institute of Aeronautics and Astronautics. (aiaawe2006-394).* http://www.personal.psu.edu/lnl/papers/aiaawe2006-394.pdf. Accessed Nov 2014.

Strømmen, E. N. (2005). *Theory of bridge aerodynamics.* USA: Springer. (ISBN: 978-3-642-13659-7).

Taghipour, R. (2008). *Efficient prediction of dynamic response for flexible and multi-body marine structures.* Norway: NTNU.

USFOS. (2014). *USFOS non-linear static and dynamic analysis of space frame structures.* http://www.usfos.no/. Accessed April 2014.

Vatne, S. R. (2011). *Aeroelastic instability and flutter for a 10 MW wind turbine.* Trondheim: NTNU.

WAMIT. (2013). *WAMIT user manual.* USA: WAMIT INC.

Zhen Gao, Nilinjan Saha, Torgeir Moan, Jorgen Amdahl (2010). Dynamic analysis of offshore fixed wind turbines under wind and wave loads using alternative computer codes. *Proceedings of the 3rd Conference on the Science of making Torque from Wind (TORQUE),* June 28–30, Heraklion, Greece.

Chapter 11
Stochastic Analyses

11.1 Introduction

Ocean environment including wave, wind and current is random in nature. Hence, the aerodynamic and hydrodynamic loads and consequently the responses of offshore structures are random. If the time variation of the loads due to random nature of metocean is neglected, then, the applied loads are called "deterministic". This means the loads and consequently dynamic responses should be exactly known or determined. Normally, time variation of the loads due to random nature of the metocean is important and should be considered in the dynamic analysis. Probabilistic and statistical approaches are developed to handle "stochastic" dynamics. The aim of such methods is to "predict" the loads and load effects as accurate as possible. During the past couple of decades, stochastic methods have been widely applied in modeling the waves, wave loads and wave-induced responses for ships and ocean structures.

Due to the strong influence of randomness of wave and wind loads on the performance and structural integrity of the offshore energy structures, nonlinear dynamics and a sophisticated aero-hydro-servo-elastic formulation considering the inertia, damping and stiffness components, the common practice is performing stochastic analysis considering probabilistic definition for both loads and responses. Keeping in mind that the cost is one of the most critical challenges for offshore energy structures, another issue is the possible cost reduction by accurate prediction of responses using stochastic methods. This can help to describe structural integrity with defined confidence intervals and avoiding overdesign or risky design.

Due to the inherent randomness in wave and wind, the responses of ocean structures subjected to similar environmental conditions can be different from one record to another. Moreover, each recorded time history is usually highly irregular. The recorded time histories look very similar, while there are significant differences between them on a local level.

© Springer International Publishing Switzerland 2014
M. Karimirad, *Offshore Energy Structures*, DOI 10.1007/978-3-319-12175-8_11

11.2   Probabilistic and Stochastic Theories

The characteristics of a random event can be described using the probability distribution of a random variable representing the phenomenon. In ocean engineering and offshore technology, statistical properties of responses occur in random fashion with respect to time. And, consideration of the time element cannot be ignored in many random phenomena. Few examples are wave profile in the ocean, response of offshore wind turbine to a wind gust, motion of a ship-shaped structure subjected to wave loads and vibration of fixed platforms caused by earthquake.

It is needed to consider a family of random variables that is a function of time in order to evaluate the statistical characteristics of random phenomena. These random phenomena are called random or "stochastic processes".

A stochastic process is a function of two arguments $\{x(t,\omega); t \in T, \omega \in \Omega\}$, where t is time and Ω is the sample space. For fixed time, $x(\omega)$ is a family of variables called an ensemble, and for a fixed ω, $x(t)$ is a function of time that is the so called sample function. Figure 11.1 shows series of sample functions: $x_1(t), x_2(t), \ldots, x_N(t)$. Two ensembles are shown as well at time t_1 and $t_1 + \tau$: $\{x_1(t_1), x_2(t_1), \ldots, x_N(t_1)\}$ and $\{x_1(t_1 + \tau), x_2(t_1 + \tau), \ldots, x_N(t_1 + \tau)\}$, respectively. So, the collection of simultaneous recorders at a particular time is an "ensemble".

The statistical properties of a stochastic process are obtained with respect to the ensemble. However, for an "ergodic" phenomenon, the statistical properties may be obtained from analysis of a single record (the ergodic properties are explained later). In Fig. 11.1, a continuous time process is shown. But, both state and time may be discrete.

- Mean value, variance, covariance

First, some of the important parameters are defined based on the probability theory. For a random variable (X), the relation between the cumulative distribution function $F(x)$ and probability density function $f(x)$ is: $f(x) = \dfrac{\partial F(x)}{\partial x}$ (assuming $F(x)$ is differentiable). The mean or expected value is defined by:

$$\mu = [EX] = \int x f(x) dx. \tag{11.1}$$

The finite sum can be used for discrete random variables; it is possible to derive the mean value by: $\mu = \sum_{i=1}^{n} x^{(i)} p_i$, in which $p_i = \Pr\{X = x^{(i)}\}$.

More conveniently, if the stochastic experiment repeats for an "infinite" number of times, it is possible to express the mean value as: $\mu = \lim_{N \to \infty} \dfrac{1}{N} \sum_{j=1}^{N} x_j$.

Consider N experiments with outcomes $x_1, x_2, \ldots, x_N (N \gg n)$. As X can assume only the $x^{(1)}, x^{(2)}, \ldots, x^{(n)}$, the outcomes are divided to n groups. $x^{(i)}$ represents the outcomes in group no. i, and N_i denotes the number of outcomes in this group.

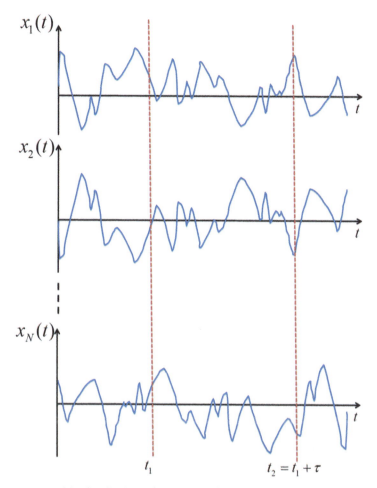

Fig. 11.1 An ensemble of realizations of a stochastic phenomenon, i.e. ocean wave elevation

Then, it is possible to show the relation between the two approaches presented above for calculation of the mean value as (Naess and Moan 2013):

$$\frac{1}{N}\sum_{j=1}^{N} x_j = \sum_{i=1}^{n} x^{(i)} \frac{N_i}{N} \quad \text{and} \quad p_i = \frac{N_i}{N}. \tag{11.2}$$

Variance of a process $Var[X]$ is defined as: $Var[X] = \sigma^2 E\left[(X-\mu)^2\right] = \int (x-\mu)^2 f(x)dx$. σ is the standard deviation (STD) of the phenomena. The coefficient of variation (COV) is defined by: σ/μ and covariance of X and Y is defined as:

$$C_{XY} = Cov[X,Y] = E[(X - E[X])(Y - E[Y])] = E[(X - \mu_X)(Y - \mu_Y)]. \tag{11.3}$$

The correlation coefficient is another useful statistical parameter defined by: $\rho_{XY} = C_{XY}/\sigma_X\sigma_Y$ (Newland 2005).

- Stationary process

As mentioned earlier, ensemble is the base for calculation of statistical properties of a random process. A process is called "stationary" if the statistical properties of a process are invariant under translation of time. Mathematically, if the joint distribution of the N-dimensional random vectors $\{x_1(t), x_2(t), \ldots, x_N(t)\}$ and $\{x_1(t+\tau), x_2(t+\tau), \ldots, x_N(t+\tau)\}$ is the same for all τ, the stochastic process is called stationary or "steady-state" stochastic phenomenon. A stochastic process satisfying this condition is also called a "strictly" stationary stochastic process. Usually, a more relaxed condition is applied for analysis of random data which is the so called weakly stationary. To define a weakly stationary stochastic process, the auto-covariance (C_{XX}) function is defined. The auto-covariance function depends on time t_1 and t_2.

$$C_{XX} = Cov\left[X(t_1), X(t_2)\right] = E\left[\left(X - E\left[X(t_1)\right]\right)\left(Y - E\left[X(t_2)\right]\right)\right] \quad (11.4)$$

If the auto-covariance is just a function of time difference (τ), see Fig. 11.1, then it is possible to write: $R(\tau) = Cov[X(t), X(t+\tau)]$ A stochastic process is weakly stationary, or covariance stationary, if its mean value is constant, independent of time, and its auto-covariance just depends on the time difference.

- Ergodic process

A stochastic process is ergodic if the time average of a single realization is "approximately" equal to the ensemble average. This means the average of each ensemble can be replaced by the average of a single record. Details of an ergodic theorem are out of scope of this book.

- Narrowband process

For continuous-state and continuous-time random processes, the stochastic process is called narrow banded if the amplitude (X) and phase (α) change slowly and randomly, while the frequency (ω_0) is a constant value as shown in Fig. 11.2. The narrowband process can be expressed as $X(t)\sin(\omega_0 t + \alpha(t))$. The stochastic process records shown in Fig. 11.1 are wideband.

- Gaussian (normal) process

A stochastic process is called Gaussian (or Normal) if for any given time its value is normally distributed (the stochastic process should satisfy the ergodic property). Several random phenomena can be represented by a normal process, and the normal process plays an important role in stochastic analysis of random phenomena, for example, wind-generated waves observed in the ocean. The wave elevation can be assumed ergodic Gaussian process and represented by the normal probability den-

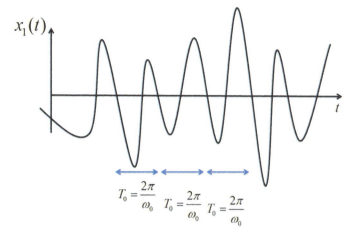

Fig. 11.2 An example of a narrowband stochastic phenomenon

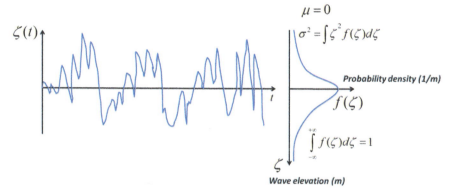

Fig. 11.3 Wave elevation time history and normal (Gaussian) distribution

sity, see Fig. 11.3. Also, the wave elevation is assumed stationary within practical time limits, i.e. 3 h (Roberts and Spanos 1990).

• Rayleigh distribution

For a broad-banded process, the extremes (individual maxima) of the Gaussian process are described by the Rice distribution. If the Gaussian process is narrowband, the Rayleigh distribution is applied to describe the statistics of extremes. For a narrowband process, there is one positive peak for each zero up-crossing. However, for a broadband process, there may be a negative maximum and positive minimum for a zero up-crossing, see Fig. 11.4. Rayleigh distribution (probability density function, PDF) is defined as:

$$f(\zeta_a) = \frac{\zeta_a}{\sigma_{\zeta_a}^2} \exp\left(-\frac{\zeta_a^2}{2\sigma_{\zeta_a}^2}\right). \tag{11.5}$$

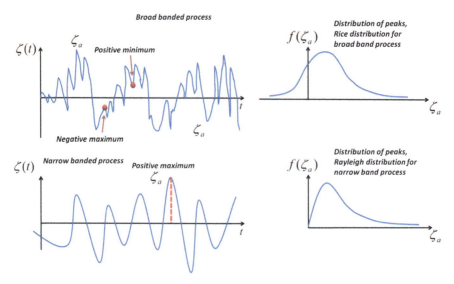

Fig. 11.4 Peaks and distribution of maxima for broad-banded and narrow-banded random processes

The formula given by Rice consists of two terms, one which has a Gaussian character with zero as a mean, and one which has the character of a Rayleigh-like distribution. The Rayleigh distribution is obtained when there are exactly two zero-crossings per peak. A Gaussian distribution is obtained when there are many peaks per zero-crossing (Bvoch 1963).

11.3   Spectrum and Spectral Analysis

Spectra can be applied to describe the stochastic processes in a frequency domain. A spectrum provides statistical information of the process respect to frequencies. Using a spectrum, the statistical information can be obtained, and it is very useful to transform time-domain results to frequency-domain spectra to understand more easily the important frequency components and dynamics involved in a physical phenomenon.

A sample time record, i.e. a response time series obtained from experiments or numerical simulation, can provide useful information and statistical data such as the process variance and mean value; the bandwidth can be calculated from it. However, several time records are needed to correctly estimate these parameters. The common practice, in offshore engineering, is to repeat numerical simulations and experiments to obtain the desired level of accuracy. Statistical information from different realizations can be averaged. Also, the spectra can be averaged to represent more realistic data. Basically, time domain simulations are post-processed to define

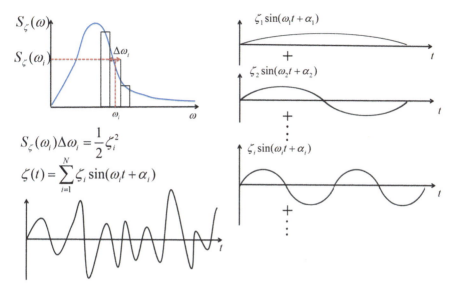

Fig. 11.5 Wave spectrum, irregular wave and its relation to regular harmonic waves and spectrum applying superposition

statistical characteristics, and they are transformed to frequency domain using numerical methods such as fast Fourier transform (FFT).

Figure 11.5 shows an irregular wave, a wave spectrum and set of regular waves and the relations between them. For an irregular ocean wave, the wave elevation at each time step can be presented by superposition of regular waves. The time histories for pressure, velocity and acceleration can be found at any time at any point by using the linear wave theory and linear superposition if the amplitudes and phase angles are found for all harmonic components in the irregular wave time history.

The spectrum moments describe the stochastic process characteristics and statistical parameters. The nth order moment of an energy spectrum is defined by the following integral:

$$m_n = \int_0^\infty S(\omega)\omega^n d\omega. \tag{11.6}$$

One of the most interesting results obtained from spectrum moments is the first moment which is linked to the STD. The area under the energy spectrum is the variance of the process:

$$m_0 = \int_0^\infty S(\omega)d\omega = Area = \sigma^2. \tag{11.7}$$

Note: The variance (σ^2) is the square of the STD. Also, the root mean square (RMS) is equal to the STD if the mean value is zero. Usually, the mean value of the dynamic responses is subtracted from the time series before calculating the STD. This makes it easier to present the statistical data of offshore energy structures subjected to wave

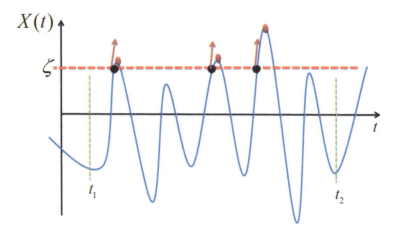

Fig. 11.6 Up-crossings of a level

and wind loads, as the wind and higher-order wave loads results in drift motions. For example, consider a floating wind turbine subjected to wind loads in a calm sea. The wind loads result in drift surge and pitch motions (socalled tilt).

There is a term called significant wave height (H_S), which is one of the most important used parameter in the field of offshore engineering. This parameter is used as an input to define the sea state spectra. Significant wave height $\left(H_S \equiv H_{1/3}\right)$ is the average value of the 1/3 largest waves in a sample. The significant wave height is linked to statistical data of the sample and the spectra moment through the following expression: $H_S = 4\sigma$.

Another important parameter is the bandwidth parameter (ε), which is used to express how much a process is broad banded or narrow banded. The bandwidth parameter can be calculated using time series or the spectrum of the process. If the bandwidth parameter is zero, the process is narrow banded. For the broad-banded stochastic process like "white noise", the bandwidth parameter is 1. The bandwidth parameter is defined by:

$$\varepsilon = \sqrt{1 - \frac{m_2^2}{m_0 m_4}}, \tag{11.8}$$

in which m_0, m_2, m_4 are moments of the spectrum, or by:

$$\varepsilon = \sqrt{1 - \left(\frac{N_{0^+}}{N_{max}}\right)^2}, \tag{11.9}$$

in which N_{0^+} is the number of zero up-crossing, and N_{max} is the total number of maxima. These data can be obtained using time series. For the narrowband process, as it is shown in Fig. 11.6, the zero up-crossing (N_{0^+}) is equal to the total number of maxima. For example, in this example for a defined time period $t_2 - t_1$, there are five zero up-crossings, and, correspondingly, there are five maxima. When the up-crossing level increases, the up-crossing rate decreases, e.g. in Fig. 11.6, there are just three up-crossings at level ζ.

The up-crossing rate ($v^+(\zeta)$), the expected number of positive crossings per time, of a stationary stochastic process at $X(t) = \zeta$ is defined by the following expression:

$$v^+(\zeta) = \int_0^\infty \dot{x} f_{X\dot{X}}(\zeta, \dot{x}) d\dot{x}, \tag{11.10}$$

in which $\dot{X} = \dfrac{\partial X(t)}{\partial t}$, and $f_{X\dot{X}}$ is the joint probability density function of X and $\dot{X}$. The zero up-crossing rate ($v^+(0)$) is obtained by $\zeta = 0$, hence:

$$v^+(0) = \int_0^\infty \dot{x} f_{X\dot{X}}(0, \dot{x}) d\dot{x} \tag{11.11}$$

For a Gaussian process (since $X(t)$ is stationary), the process and its derivative are statistically independent. So, the joint probability density function is defined as:

$$f_{X\dot{X}}(x, \dot{x}) = f_X(x) f_{\dot{X}}(\dot{x}) \tag{11.12}$$

$$f_X(x) = \frac{1}{\sqrt{2\pi}\sigma_X} \exp\left(-\frac{x^2}{2\sigma_X^2}\right) \quad \text{and} \quad f_{\dot{X}}(\dot{x}) = \frac{1}{\sqrt{2\pi}\sigma_{\dot{X}}} \exp\left(-\frac{\dot{x}^2}{2\sigma_{\dot{X}}^2}\right). \tag{11.13}$$

Hence, for a Gaussian process with zero mean value, the up-crossing can be derived as:

$$v^+(\zeta) = \frac{1}{2\pi} \frac{\sigma_{\dot{X}}}{\sigma_X} \exp\left(-\frac{\zeta^2}{2\sigma_X^2}\right) \tag{11.14}$$

$$v^+(0) = \frac{1}{2\pi} \frac{\sigma_{\dot{X}}}{\sigma_X} = \frac{1}{2\pi} \sqrt{\frac{m_2}{m_0}} \tag{11.15}$$

$$v^+(\zeta) = v^+(0) \exp\left(-\frac{\zeta^2}{2\sigma_X^2}\right). \tag{11.16}$$

11.4 Peaks Distribution for General Random Process: Rice Distribution

As it has been mentioned in the previous part, the Rayleigh distribution can be used to present the maxima distribution for a narrow-banded stochastic process. Rice distribution gives the distribution of maxima (all positive and negative maxima) for a general stationary Gaussian random process having zero mean value (Cartwright and Longuet-Higgins 1956 and Bvoch 1963).

$$f(\zeta_a) = \frac{1}{\sigma_{\zeta_a}\sqrt{2\pi}}\,\varepsilon\exp\left(-\frac{\zeta_a^2}{2\varepsilon^2\sigma_{\zeta_a}^2}\right) + \sqrt{1-\varepsilon^2}\,\frac{\zeta_a}{\sigma_{\zeta_a}^2}\exp\left(-\frac{\zeta_a^2}{2\sigma_{\zeta_a}^2}\right)\phi\left(\frac{\zeta_a}{\varepsilon\sigma_{\zeta_a}}\sqrt{1-\varepsilon^2}\right) \quad (11.17)$$

ϕ is the cumulative Gaussian distribution specified by:

$$\phi\left(\frac{\zeta_a}{\varepsilon\sigma_{\zeta_a}}\sqrt{1-\varepsilon^2}\right) = \frac{1}{\sqrt{2\pi}}\int_{-\infty}^{\frac{\zeta_a}{\varepsilon\sigma_{\zeta_a}}\sqrt{1-\varepsilon^2}}\exp\left(-\frac{t^2}{2}\right)dt. \quad (11.18)$$

The expected frequency between all maxima (both positive and negative maxima) is given by:

$$\upsilon_{max} = \frac{1}{2\pi}\sqrt{\frac{m_4}{m_2}}. \quad (11.19)$$

The Rice distribution covers the Gaussian and Rayleigh distributions.

$$\varepsilon = 1 \Rightarrow f(\zeta_a) = \frac{1}{\sqrt{2\pi}\sigma_{\zeta_a}}\exp\left(\frac{\zeta_a^2}{2\sigma_{\zeta_a}^2}\right) \quad \text{Gaussian}$$

$$\varepsilon = 0 \Rightarrow f(\zeta_a) = \frac{\zeta_a}{\sigma_{\zeta_a}^2}\exp\left(-\frac{\zeta_a^2}{2\sigma_{\zeta_a}^2}\right) \quad \text{Rayleigh} \quad (11.20)$$

For the narrow-banded process, the bandwidth parameter is zero, hence:

$$m_2^2 = m_0 m_4 \Leftrightarrow \upsilon_{max} = \upsilon_{0^+} = \frac{1}{2\pi}\sqrt{\frac{m_2}{m_0}}. \quad (11.21)$$

The Rayleigh distribution gives an upper limit for the distribution of maxima. Hence, if the Rayleigh distribution is applied to define the probability of exceedance of a level, the estimation is conservative compared to the case when estimation is obtained by applying the Rice distribution (Fig.11.7).

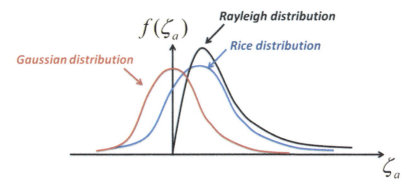

Fig. 11.7 Gaussian, Rayleigh and Rice distribution. Rayleigh represents maxima for narrow-banded processes, and Rice distribution is used for maxima of wide-banded processes; Rayleigh and Gaussian functions are special cases of the Rice function

11.5   1/N Largest Maxima

The maximum value (ζ_{max}) which is exceeded by the probability of $1/N$ is interesting for design purposes of offshore structures.

$$P(\zeta_{max} > \zeta_{1/N}) = 1/N = \int_{\zeta_{1/N}}^{\infty} f(\zeta_a)d\zeta_a \qquad (11.22)$$

For Rayleigh distribution, the integration gives $\zeta_{1/N} = \sigma\sqrt{2\ln N}$.

The mean value of the $1/N$ largest maxima, $\overline{\zeta}_{1/N} = E[\zeta_{1/N}]$, which is the centre of the shaded area in Fig. 11.8, is defined by:

$$\overline{\zeta}_{1/N} = E[\zeta_{1/N}] = \sigma\sqrt{2\ln N} + \sigma N\sqrt{\frac{\pi}{2}}[1 - erf(\sqrt{\ln N})] \qquad (11.23)$$

The significant wave amplitude ($\overline{\zeta}_{1/3}$) is $\overline{\zeta}_{1/3} = \sigma\sqrt{2\ln 3} + 3\sigma\sqrt{\frac{\pi}{2}}1 - erf(\sqrt{\ln 3}) \approx 2\sigma$. Hence, the significant wave height (H_S) is defined by $H_{1/3} = 4\sigma$. This means that the significant wave height expression is valid for wave heights which are Rayleigh distributed.

11.6   Largest Maximum Among N Maxima

A simple way to check the structural integrity of an offshore structure subjected to wave loads is to consider a large wave. This is a very simple and preliminary check. To perform such an analysis, a proper estimation of the largest wave height is

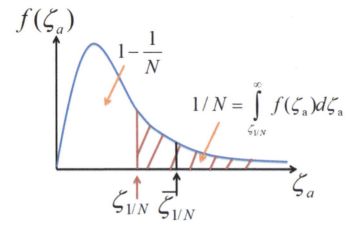

Fig. 11.8 $1/N$ largest maxima and its relation to the probability density function

needed. Spectral and statistical approaches are used to define such a "design wave". The "largest maximum among N maxima" is described by defining the extreme value distribution.

If N maxima denoted by $\zeta_{\max 1}, \zeta_{\max 2}, \ldots, \zeta_{\max N}$ happens in a time interval, and the largest maximum of them is denoted by ζ_{LM}, the distribution of ζ_{LM} can be defined as explained below.

$$\mathrm{P}(\zeta_{LM} \leq \zeta) = P(all\, \zeta_{\max i} \leq \zeta) = P\left[(\zeta_{\max 1} \leq \zeta) \cap (\zeta_{\max 1} \leq \zeta) \cap \ldots \cap (\zeta_{\max N} \leq \zeta)\right] \quad (11.24)$$

It is assumed that all the maxima are Rayleigh distributed, and all of them are independent. Hence, the probability of ζ_{LM} being less than a certain level is:

$$\mathrm{P}(\zeta_{LM} \leq \zeta) = P(\zeta_{\max 1} \leq \zeta).P(\zeta_{\max 1} \leq \zeta) \ldots P(\zeta_{\max N} \leq \zeta). \qquad (11.25)$$

$P(\zeta_{\max i} \leq \zeta)$ is the cumulative distribution function of the individual maxima. Rayleigh cumulative distribution function (CDF) and PDF are:

$$F(\zeta_a) = 1 - \exp\left(-\frac{\zeta_a^2}{2\sigma_{\zeta_a}^2}\right)$$

$$f(\zeta_a) = \frac{\partial F(\zeta_a)}{\partial \zeta_a} \Rightarrow f(\zeta_a) = \frac{\zeta_a}{\sigma_{\zeta_a}^2} \exp\left(-\frac{\zeta_a^2}{2\sigma_{\zeta_a}^2}\right). \qquad (11.26)$$

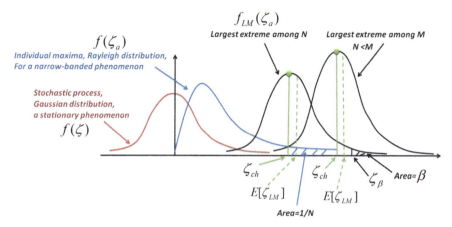

Fig. 11.9 Largest maximum among N maxima

Therefore, the cumulative distribution of the largest maxima is written as follows:

$$F_{LM}\left(\zeta_a\right) = \left[F\left(\zeta_a\right)\right]^N = \left[1 - \exp\left(-\frac{\zeta_a^2}{2\sigma_{\zeta_a}^2}\right)\right]^N$$

$$f_{LM}\left(\zeta_a\right) = N\left[F\left(\zeta_a\right)\right]^{N-1} f\left(\zeta_a\right) = N\left[1 - \exp\left(-\frac{\zeta_a^2}{2\sigma_{\zeta_a}^2}\right)\right]^{N-1} \frac{\zeta_a}{\sigma_{\zeta_a}^2}\exp\left(-\frac{\zeta_a^2}{2\sigma_{\zeta_a}^2}\right).$$

$$(11.27)$$

The expected value of the largest maximum can be calculated by:

$$E\left[\zeta_{LM}\right] = \int_0^{\infty} \zeta_a f_{LM}\left(\zeta_a\right)d\zeta_a \cong \sigma_{\zeta_a}\left(\sqrt{2\ln N} + \frac{0.57722}{\sqrt{2\ln N}} + O\left((\ln N)^{-1.5}\right)\right). \quad (11.28)$$

For large N, the error ($O((\ln N)^{-1.5})$) is negligible. The most probable maximum, the socalled characteristic largest value is another important parameter which is usually applied. It is the corresponding value when $f_{LM}\left(\zeta_a\right)$ is a maximum.

$$\left.\frac{\partial f_{LM}\left(\zeta_a\right)}{\partial \zeta_a}\right|_{\zeta_a=\zeta_{ch}} = 0 \Rightarrow \zeta_{ch} \cong \sigma_{\zeta_a}\sqrt{2\ln N} \qquad (11.29)$$

Figure 11.9 illustrates the probability density functions for process, individual maximum and largest maximum. When N increases, the probability density function moves toward right, i.e. $M > N$ in the current example shown in the figure. As clearly shown in Fig. 11.9, the probability that a maximum is larger than the "characteristic largest value" is high.

$$P\left(\zeta_{LM} > \zeta_{ch}\right) = 1 - P\left(\zeta_{LM} \leq \zeta_{ch}\right) = 1 - F_{LM}\left(\zeta_{ch}\right)$$

$$= 1 - \left[F_{\zeta_a}\left(\zeta_{ch}\right)\right]^N = 1 - \left[1 - \exp\left(-\frac{\zeta_{ch}^2}{2\sigma_{\zeta_a}^2}\right)\right]^N$$

$$\cong 1 - \left[1 - \exp\left(-\frac{\sigma_{\zeta_a}^2 2\ln N}{2\sigma_{\zeta_a}^2}\right)\right]^N$$

$$= 1 - [1 - \exp(-\ln N)]^N = 1 - \left[1 - \frac{1}{N}\right]^N \approx 1 - \frac{1}{e} = 0.63. \qquad (11.30)$$

We are interested to find the largest maximum exceeded by the probability of β, see Fig. 11.9 in which β is a small number. This results in a larger extreme value, which can be used as a reference design value.

$$P\left(\zeta_{LM} > \zeta_\beta\right) = \beta = 1 - F_{LM}\left(\zeta_\beta\right) = 1 - \left[F_{\zeta_a}\left(\zeta_\beta\right)\right]^N$$

$$\left[F_{\zeta_a}\left(\zeta_\beta\right)\right]^N = 1 - \beta \rightarrow \left[1 - \exp\left(-\frac{\zeta_\beta^2}{2\sigma_{\zeta_a}^2}\right)\right]^N = 1 - \beta$$

$$1 - \exp\left(-\frac{\zeta_\beta^2}{2\sigma_{\zeta_a}^2}\right) = (1 - \beta)^{1/N} \approx 1 - \frac{\beta}{N} \Rightarrow \zeta_\beta = \sigma_{\zeta_a}\sqrt{2\ln\frac{N}{\beta}} \qquad (11.31)$$

11.7   Extreme Value Analysis

These are three classical types of extreme value distributions: Gumbel, Frechet and Weibull. For instance, Gumbel distribution for extreme values is presented by:

$$F^{Gum}\left(\zeta_a\right) = \exp\left(-\exp\left(-\alpha(\zeta_a - u)\right)\right)$$

$$f^{Gum}\left(\zeta_a\right) = \alpha\exp\left(-\alpha(\zeta_a - u) - \exp\left(-\alpha(\zeta_a - u)\right)\right). \qquad (11.32)$$

The Gumbel parameters can be found using the initial distribution, e.g. by knowing the peak value distribution for wave elevation, $F(\zeta_a), f(\zeta_a)$, as follows:

$$F(u) = 1 - \frac{1}{N}$$

$$\alpha = Nf(u). \qquad (11.33)$$

Mean value for Gumbel distribution is defined as:

$$\mu^{Gum} = u + \frac{0.57722}{\alpha}. \tag{11.34}$$

Starting with Rayleigh distribution, we can find the mean value ($F^{Gum}(\zeta_a) = 0.5$) for the largest maximum among N maxima:

$$F(u) = 1 - \exp\left(-\frac{u^2}{2\sigma_{\zeta_a}^2}\right) = 1 - \frac{1}{N}$$

$$u = \sigma_{\zeta_a} \sqrt{2\ln N}$$

$$\alpha = \frac{\sqrt{2\ln N}}{\sigma_{\zeta_a}} \tag{11.35}$$

$$\mu^{Gum} = u + \frac{0.57722}{\alpha} = \sigma_{\zeta_a}\left(\sqrt{2\ln N} + \frac{0.57722}{\sqrt{2\ln N}}\right). \tag{11.36}$$

The most probable largest extreme (when $f^{Gum}(\zeta_a)$ is maximum) is $\sigma_{\zeta_a}\sqrt{2\ln N}$.

Applying classical methods for predicting the extreme values requires comprehensive effort to determine the type of the extreme value distribution and its parameters which can be uncertain. The uncertainty increases for offshore energy structures, as limited data for extreme values of these structures are available. However, for land-based wind turbines, various techniques for the estimation of extreme loads/responses have been used. Peak over threshold methods using Weibull models and block maxima techniques are some of the attempts to model the extreme value statistics by determining the extreme value distribution.

Several methods such as Monte Carlo methods, the Weibull tail, the Gumbel method, the Winterstein method and the peaks-over-threshold method (POT) are presented to estimate the extreme value. The analytical models are used for determining the linear response while the distribution of the nonlinear response generally needs to be treated in a semiempirical manner by modeling the distribution of the response peaks or up-crossing rates (Karimirad 2011).

Extreme value statistics for a 1 or 3-h period may be obtained by taking into account the regularity of the tail region of the mean up-crossing rate. The prediction of low-exceedance probabilities needs a large sample size which results in time-demanding calculations, as extreme values have a low probability of occurrence. Also, the analyses of offshore energy structures subjected to stochastic wave and wind loading are time consuming. For floating concepts, this computational cost is even worse because the total simulation time is higher. So, extrapolation methods can be used to estimate the extreme value responses of these structures.

The mean up-crossing rate can be implemented for extreme value prediction. For complicated marine structures subjected to wave and wind loads, i.e. offshore energy structures, the response is nonlinear and non-Gaussian. Consequently, the methods based on the up-crossing rate are more robust and accurate. The up-crossing rate is the frequency of passing a specified response level (up-crossing rate is lower for higher response levels). The Poisson distribution represents the extreme values since the occurrence of extreme values is rare. Also, the Poisson distribution can be defined based on the up-crossing rate. For each response level, it is possible to count the number of up-crossings directly from the time histories. Long time domain simulations are needed to obtain up-crossing rates for high response levels. For example, a 1-h simulation cannot provide any information about an up-crossing rate of 0.0001. Extrapolation methods are applied to extrapolate raw data and provide up-crossing rates for higher response levels (Naess et al. 2008). Probability of extreme values using Poisson distribution can be written as:

$$P(X(T) \le y) = \exp\left(-\int_0^T \upsilon_y^+(t)dt\right),$$ (11.37)

in which T is the total time duration, i.e. 3-h, and $\upsilon_y^+(t)$ is the up-crossing rate of level y. For more information refer to Karimirad and Moan (2011).

11.8   Stochastic Time Domain Analysis Aspects

Frequency domain procedure is discussed in the previous chapter. For single degree of freedom (SDOF) systems, the frequency domain analysis is straightforward which results in several useful outputs, i.e. mechanical/structural and hydrodynamic transfer functions. The straightforward formulation outlined for SDOF cannot be applied when dealing with multi-degree-of-freedom (MDOF) systems. The reason is that the phase angle between load components for different degrees of freedom must be considered as well. Hence, the deformations are described by both auto-spectra and cross-spectra. The cross-spectra define the correlation (or phases) between responses.

Frequency domain procedure can only be applied for linear systems and Gaussian processes. This is one of the main limitations as nonlinearities, i.e. drag forces, introduce non-Gaussian processes. Normally, the stochastic linearization is applied to linearize the dynamic systems of offshore structures. The procedures for equivalent (stochastic) linearization can be found in literatures (this is not further discussed). Some of the time domain procedure aspects are described herein.

Time integration procedures for dynamic analyses can be applied for stochastic analysis. The applied loads, particularly wave and wind loads, are stochastic processes. The stochastic analysis of systems including nonlinearity is possible by using the time domain approach.

The resulting responses and load effects, i.e. bending moments and stresses in a specific cross section, are stochastic processes obtained from limited periods of time, i.e. from 3-h analyses. Hence, statistical and spectral data from distribution function, spectral moments, mean value, skewness, kurtosis, variance, extremes, etc. should be estimated using proper statistical/probabilistic methods.

Time domain simulations start by generating wave and wind input data. The time domain representation of stochastic waves is given in Fig. 11.5. The random phase angle (α) is drawn applying "pseudorandom" number generators. The quality of the garneted waves is linked to randomness of the phase angle. Hence, proper quality tests of generated wave and wind time series should be performed as some generators may have hidden correlations which generate non-Gaussian time series (Larsen 1990).

FFT and inverse fast Fourier transform (IFFT) are usually applied for stochastic dynamic analysis. The time domain results can be transformed to the frequency domain using FFT methods. The spectrum can be calculated from a recorded time series with constant sampling interval. Also, input time series can be generated from a wave or wind spectrum, see Fig. 11.5. Usually, the total time series with equally spaced time increments are generated and stored. The relation between time increments, frequency increments, number of frequencies and number of time steps are as follows:

$$N_{\Delta t} = 2N_{\Delta\omega}, \quad T = N_{\Delta t}\Delta t, \quad \omega_{max} = N_{\Delta\omega}\Delta\omega$$

$$T = 2\pi/\Delta\omega, \quad T_{min} = 2\pi/\omega_{max}. \tag{11.38}$$

For the generation of turbulent wind, if the turbulence intensity and the mean wind speed are defined for a site, the inverse discrete fourier transform (DFT) approach can be applied to develop the turbulent wind field (Hansen 2008). The Fourier transform that satisfies the wind spectrum ($S(\omega_n)$) can be presented as:

$$V(t) = \bar{V} + \sum_{n=1}^{N/2} \sqrt{\frac{2S(\omega_n)}{T}} \cos(\omega_n t - \varphi_n)$$

$$t = i \times \Delta t \ for \ i = 1, \ldots, N, \tag{11.39}$$

where T is the total time. As for wave generation, the phase angle φ_n is not reflected in the wind spectrum and should be modeled using a random number generator yielding a value between 0 and 2π. By performing a DFT of the wind spectrum, all frequencies between 0 and ∞ are involved. However, frequencies between $1/T$ and $N/2T$ are sufficient for generating proper wind time series (Karimirad 2011). For a 3D wind simulation, not only the frequency but also the distance between different points is important. This means the time series for different points are dependent. Hence, the wind velocities at different blade locations are correlated. This is introduced by the coherence function. The correlation will increase if the distance between the points decreases.

The generated wave and wind time series should not contain repetitions. As the time series are periodic, the time history may repeat after a given time $T = 2\pi/\Delta\omega_{min}$ in which $\Delta\omega_{min}$ is the smallest difference between two neighbour frequencies.

Most of the 3D turbulent wind generators have limitations in the size of the turbulence box. Hence, for some load cases especially for high wind conditions, it is needed to split the analyses to shorter simulation periods with different seeds to ensure an accurate stochastic representation of the input wave and wind data. The responses of the shorter simulation periods should properly be combined. Karimirad has applied this method to investigate the extreme responses of offshore wind turbines (Karimirad and Moan 2011). The number of the wave frequencies and cut-in and cut-out wave frequencies of the wave spectrum are connected to repetition of the waves in a defined time domain analysis. Fewer frequencies and wider frequency range may lead to a repetition of waves.

The motion equations presented in the previous chapter should be set up taking into account the coupling after proper generation of the wave and wind time series. The position-dependent aerodynamic and hydrodynamic loads are calculated based on the presented aerodynamic and hydrodynamic theories at each time step. Integration methods are used to calculate the responses step by step. At each time step, the responses including the position of the structure are found, and all the terms in equations of motions are updated by considering the geometrical and environmental conditions for the new time step.

The time domain analysis for offshore energy structures, especially for floating structures having low-frequency components, is very time demanding. To capture accurate aero-hydro-elastic dynamic responses of these structures, a sufficient simulation time is required. The lowest natural frequency of the system is important in this regard. For instance, if the surge natural period is around 100 s then each 3-h simulation contains more than 100 surge cycles which is enough to capture a correct representation of energy at this particular low-frequency response.

The accuracy of the time domain simulations is highly dependent to time steps, and large time steps are not suggested. When it comes to dynamic responses, the maximum time step should be smaller than a certain value to capture, i.e. 40 cycles of the lowest natural period involved in the system. For instance, to capture the blade eigenfrequencies which are around 3 rad/sec, 40 cycles can be achieved by choosing 0.05 s for the time step. Usually, smaller time steps are required to capture accurate aeroelastic responses. The suitable time step is dependent to concept/design, aero-hydro-elastic code, environmental conditions, system status (i.e. operational or fault conditions) as well as the specific response under consideration. The best practice is to perform a convergence study for finding the required time step.

11.9   Fatigue Damage Assessment: A Stochastic Analysis

As mentioned earlier, offshore energy structures have mechanical moving parts which are subjected to wear, fracture and fatigue. The cyclic loadings from the rotating rotor and power take-off systems introduce fatigue which is accumulated. The design life of mechanical components and structural members are highly affected by accumulated damage. Not only the normal operation but also other load cases such as start-up, shutdown, transportation/installation, fault conditions, extreme environmental

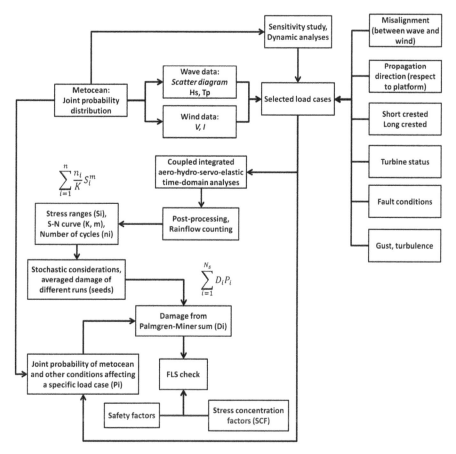

Fig. 11.10 Fatigue analysis for offshore energy structures. *FLS* Fatigue Limit State

conditions as well as abnormal conditions should be considered while assessing the fatigue life. Fatigue limit state is defined to guide fatigue design and proper estimation of fatigue life of a system and its components. Like other offshore structures, marine renewable energy devices are subjected to wave and wind loads. These loads are inherently stochastic which necessitate stochastic analysis for fatigue assessments. Fatigue life of a component is usually governed by moderate environmental conditions rather than extreme events. The reason is that the fatigue damage accumulation means repetition of loads and load effects which are more for more frequent events.

In offshore industry, frequency domain analysis is widely applied to assess the fatigue damage for oil/gas applications. This can be applied as the wave loads can be linearized with an acceptable accuracy. However, taking into account the non-linearities for offshore energy structures is required. Hence, stochastic time domain simulations are basically needed.

Fatigue damage examination is a long-term analysis, see Fig. 11.10. This means all the possible environmental conditions with probability of occurrences of them should be defined prior to analyses. An environmental condition for an offshore energy structure is defined by wave and wind characteristics including mean wind speed, turbulence intensity, significant wave height, wave peak period as well as the joint probability of occurrence of such event in a long-term perspective. It is assumed that the environmental conditions are stationary in a certain period of time, e.g. 1 or 3 h. As the stationary assumption and simulation time affect the results, it is needed to perform a proper investigation to find the period of time needed for each individual simulation. Moreover, the stochastic realizations of wave and wind time series affect the results. Hence, several simulations with different seed numbers should be carried out to average the responses. This may be in order of ten simulations for each environmental condition which highlights the processing and post-processing time needed for fatigue analysis of offshore energy structures such as floating wind turbines. Directionality of the wave and wind respect to structure as well as misalignments (angle between wave and wind) add to the complexity of fatigue analysis of Offshore Energy Structures (OES). Moreover, waves are short crested in real life which should be considered in practice.

As it is clear, thousands of simulations may be needed to run to investigate the fatigue life of an OES. This makes it impossible in some cases as integrated time domain simulations considering aero-servo-hydro-elastic for coupled wave and wind loadings are computationally demanding. Hence, proper sensitivity studies are needed to define which parameters are more important and affecting the responses (fatigue life). The conclusions are highly concept dependent, and it is not possible to make a general rule. However, these kinds of studies are useful to minimize the efforts needed for Fatigue Limit State (FLS) checks.

References

Bvoch, J. T. (1963). Effects of Spectrum Non-linearities upon the peak distribution of random signals. *Technical Review*. Lyngeby: Brue L & Kjaer.

Cartwright, D. E., & Longuet-Higgins, M. S. (1956). The statistical distribution of the maxima of a random function. *Proceedings of the Royal Society of London. Series A, Mathematical and Physical Sciences, 237*(1209), 212–232.

Hansen, M. O. (2008). *Aerodynamics of Wind Turbines*. UK: Erathscan.

Karimirad, M. (2011). *Stochastic dynamic response analysis of spar-type wind turbines with catenary or taut mooring systems*. PhD thesis, NTNU, Norway.

Karimirad, M., & Moan, T. (2011). Extreme dynamic structural response analysis of catenary moored spar wind turbine in harsh environmental conditions. *Journal of Offshore Mechanics and Arctic Engineering, 133,* 041103-1.

Naess, A., & Moan, T. (2013). *Stochastic dynamics of marine structures*. USA: Cambridge University Press.

Naess, A., Gaidai, O., & Teigen, P. S. (2008). Extreme response prediction for nonlinear floating offshore structures by Monte Carlo simulation. *Applied Ocean Research, 29*(4), 221–230.

Newland, D. E. (2005). *An introduction to random vibrations, spectral & wavelet analysis*. USA: Dover Civil and Mechanical Engineering.

Roberts, J. B., & Spanos, P. D. (1990). *Random vibration and statistical linearization*. UK: Wiley.

Short Biography

Dr. Madjid Karimirad is a Scientist in the field of marine structures and offshore technology with almost 10 years of work and research experience. Karimirad started his career as a project-engineer after finishing BSc. He got his PhD (2011) in marine structures, his MSc (2007) in mechanical engineering/ships structures and his BSc (2005) in mechanical/marine engineering. Dr. Karimirad has served as post-doctoral academic staff; also, he has worked in industry as senior-engineer. Offshore energy structures, marine structures as well as offshore oil and gas technologies are among his research and work interests; Dr. Karimirad has carried out different projects focusing on these issues during the past years. His knowledge covers several aspects of offshore mechanics, hydrodynamics, and structural engineering. The work and research results have been published in several technical reports, theses, book chapters, journal scientific papers, and conference proceedings.

© Springer International Publishing Switzerland 2014
M. Karimirad, *Offshore Energy Structures,* DOI 10.1007/978-3-319-12175-8

Bibliography

Chapter 2

Anderson, P. (2007). SD8418: One hundred and twenty feet up! http://www.geograph.org.uk/photo/632554. Accessed Feb 2014.

Cryan, P. (2013). Wind turbine blade. http://gallery.usgs.gov/photos/10_19_2009_s84Aq11PPk_10_19_2009_3. Accessed Feb 2014.

mobiusinstitute.com. (n.d.). Vibration analysis of wind turbines. http://www.mobiusinstitute.com/articles.aspx?id=873. Accessed Feb 2014.

Ruviaro, M., Rüncos, F., Sadowski, N., & Borges, I. M. (2012). Analysis and test results of a brushless doubly fed induction machine with rotary transformer. *IEEE Transactions on Industrial Electronics, 59*(6), 2670–2677.

openei.org. (2010). Vestas. http://en.openei.org/wiki/. Accessed Feb 2014.

Chapter 3

4C-Offshore. (2013a). WindFloat. http://www.4coffshore.com/windfarms/windfloat--phase-1-portugal-pt01.html. Accessed April 2014.

ABB. (2013). ABB wind turbine converters. www.abb.com/windpower. Accessed April 2014.

Anderson, D.C., Whale, J., Livingston, P.O., & Chan, D. (2008). Rooftop wind resource assessment using a three dimensional ultrasonic anemometer. http://www.ontario-sea.org/Storage/26/1798_A_Wind_Resource_Assessment_on_a_Rooftop_Using__3D_Ultrasonic_Anemometer.pdf. Accessed April 2014.

Anderson, P. (2007). *SD8418:* One hundred and twenty feet up! http://www.geograph.org.uk/photo/632554. Accessed April 2014.

Anderson, P. (2008a). Gearbox, rotor shaft and disk brake assembly. http://www.geograph.org.uk/. Accessed April 2014.

Anderson, P. (2008b). Interior of the hub of turbine no 3. http://www.geograph.org.uk/photo/754033. Accessed April 2014.

Bauman, R. P. (2007). The Bernoulli Conundrum. http://www.introphysics.info. Accessed April 2014.

BBC. (2011). Region's MPs lobbied over Dogger Bank wind farm. http://www.bbc.co.uk/news/uk-england-15274152. Accessed April 2014.

BBC-News-Technology. (2010). Thanet offshore wind farm starts electricity production. http://www.bbc.co.uk/news/technology-11395824. Accessed April 2014.

belwind.eu. (n.d.). Offshore-wind-farm. http://www.belwind.eu/en/offshore-wind-farm/. Accessed April 2014.

bine.info. (2009). RAVE—research on the offshore test field. http://www.bine.info/en/publica-tions/publikation/rave-forschen-am-offshore-testfeld/?artikel=2214. Accessed April 2014.

blekingeoffshore.se. (2013). Blekinge Offshore. http://www.blekingeoffshore.se/. Accessed April 2014.

Carlin, P. W., Laxson, A. S., & Muljadi, E. B. (2003). The history and state of the art of variable-speed wind turbine technology. *Wind Energy, 6,* 129–159. doi:10.1002/we.77. Accessed April 2014.

Cryan, P. (2013). Wind turbine blade. http://gallery.usgs.gov/photos/10_19_2009_s84Aq11PPk_10_19_2009_3. Accessed April 2014.

demowfloat.eu. (n.d.). WindFloat. http://www.demowfloat.eu/. Accessed April 2014.

dongenergy.com. (n.d.). About gunfleet sands. http://www.dongenergy.com/Gunfleetsands/GunfleetSands/AboutGFS/Pages/default.aspx. Accessed April 2014.

eastangliawind.com. (2009–2013). East Anglia offshore windfarm zone. http://www.eastangliawind.com/. Accessed April 2014.

ecoseed.org. (2011). South Korea to build $9 billion offshore wind farm by 2019. http://www.ecoseed.org/business/asia/article/130-asia/11823-south-korea-to-build-$9-billion-offshore-wind-farm-by-2019. Accessed April 2014.

Enerpac. (2009). Perpendicular at Open Sea. http://www.ynfpublishers.com/2009/09/perpendicular-at-open-sea/. Accessed April 2014.

KarleHorn. (2010). Windkraftanlage Rotorblatt Achse. de.wikipedia.

Lorc. (2011a). Hywind Demonstration Offshore Wind Farm. http://www.lorc.dk/offshore-wind-farms-map/hywind-demonstration. Accessed April 2014.

Lorc-website. (2011). http://www.lorc.dk/. Accessed April 2014.

Lorc. (2011b). Knowledge. http://www.lorc.dk/. Accessed April 2014.

Maciel, J. G. (2010). The windfloat project. http://www.wavec.org/content/files/11_Joao_Maciel.pdf. Accessed April 2014.

marubeni.com. (2013). Fukushima recovery, experimental offshore floating wind farm project. (Nov 11) http://www.marubeni.com/news/2013/release/fukushima e.pdf. Accessed April 2014.

mhi-global.com. (2013). MHI news. http://www.mhi-global.com/discover/graph/news/no171.html. Accessed April 2014.

mobiusinstitute.com. (n.d.). Vibration analysis of wind turbines. http://www.mobiusinstitute.com/articles.aspx?id= 873. Accessed April 2014.

Muljadi, E., & Butterfield, C. P. (1999). Pitch-controlled variable-speed. Presented at the 1999 IEEE Industry Applications. Phoenix, Arizona: NREL/CP-500–27143.

nexans.com. (n.d.). Hywind—floating windpower. http://www.nexans.com/eservice/Corporate-en/navigate_252193/Hywind_floating_windpower.html. Accessed April 2014.

npd.no. (2012). Trollet som ble temmet. http://www.npd.no/Templates/OD/Article.aspx?id=4438. Accessed April 2014.

openei.org. (2010). Vestas. http://en.openei.org/wiki/. Accessed April 2014.

principlepowerinc.com. (n.d.). Principlepower. http://www.principlepowerinc.com/. Accessed April 2014.

renewablesinternational.net. (2013). First floating offshore wind substation underway. http://www.renewablesinternational.net/first-floating-offshore-wind-substation-under-way/150/435/71904/. Accessed April 2014.

rt.com. (2013). Japan to start building world's biggest offshore wind farm this summer. http://rt.com/news/japan-renewable-energy-resource-290/. Accessed April 2014.

Schlipf, D. (2012). Lidar assisted control of wind turbines. http://rasei.colorado.edu/wind-research-internal1/David_Schlipf.pdf. Accessed April 2014.

seas-nve.dk. (n.d.). Vindeby offshore wind farm. http://www.seas-nve.dk/AboutSeasNve/Wind/References/Offshore/Vindeby.aspx. Accessed April 2014.

siemens.com. (2009a). Hywind: Siemens and StatoilHydro install first floating wind turbine. http://www.siemens.com/press/en/pressrelease/?press=/en/pressrelease/2009/renewable_energy/ere200906064.htm. Accessed April 2014.

siemens.com. (2009b). Siemens wind turbine SWT-2.3-82 VS. http://www.energy.siemens.com/hq/pool/hq/power-generation/wind-power/E50001-W310-A123-X-4A00_WS_SWT-2.3-82%20VS_US.pdf. Accessed April 2014.

statoil.com. (2009). Hywind—the world's first full-scale floating wind turbine. http://www.statoil.com/en/technologyinnovation/newenergy/renewablepowerproduction/offshore/hywind/pages/hywindputtingwindpowertothetest.aspx. Accessed April 2014.

statoil.com. (2012). Hywind. http://www.statoil.com/no/technologyinnovation/newenergy/renewablepowerproduction/offshore/hywind/pages/hywindputtingwindpowertothetest.aspx. Accessed April 2014.

Statoil-Statkraft. (n.d.). Sheringham Shoal Website. http://www.scira.co.uk/. Accessed April 2014.

Technip. (2009). Technip joins StatoilHydro in announcing the inauguration of the world's first full-scale floating wind turbine. http://www.technip.com/en/press/technip-joins-statoilhydro-announcing-inauguration-worlds-first-full-scale-floating-wind-turbi. Accessed April 2014.

The-Hywind-O & M-Team. (2012). Hywind: Two years in operation. http://www.sintef.no/project/Deepwind%202012/Deepwind%20presentations%202012/D/Trollnes_S.pdf. Accessed April 2014.

VDN. (2007). TransmissionCode 2007. Berlin: Verband der Netzbetreiber—VDN—e.V. beim VDEW.

Vidigal, A. (2012). WindFloat: Quase há um ano no Mar. http://www.antoniovidigal.com/drupal/cd/WindFloat-Quase-h%C3%A1-um-ano-no-Mar. Accessed April 2014.

windpowermonthly.com. (2012). Close up—Envision's 3.6 MW offshore machine. http://www.windpowermonthly.com/article/1115270/close--envisions-36mw-offshore-machine. Accessed April 2014.

xodusgroup.com. (2013). Hywind Scotland Pilot Park Project. Edinurgh: A-100142-S00-REPT–001.

Chapter 4

ABB. (2013). ABB wind turbine converters. www.abb.com/windpower. Accessed March 2014.

alpha-ventus.de. (2010). alpha ventus—the first German offshore wind farm. http://www.alpha-ventus.de/index.php?id=120. Accessed March 2014.

Anderson, P. (2007). SD8418: One hundred and twenty feet up! http://www.geograph.org.uk/photo/632554. Accessed March 2014.

Anderson, P. (2008a). Gearbox, rotor shaft and disk brake assembly. http://www.geograph.org.uk/. Accessed March 2014.

Anderson, P. (2008b). Interior of the hub of Turbine No 3. http://www.geograph.org.uk/photo/754033. Accessed March 2014.

Anderson, D. C., Whale, J., Livingston, P. O., & Chan, D. (2008). Rooftop wind resource assessment using a three dimensional ultrasonic anemometer. http://www.ontario-sea.org/Storage/26/1798_A_Wind_Resource_Assessment_on_a_Rooftop_Using__3D_Ultrasonic_Anemometer.pdf. Accessed March 2014.

bard-offshore.de. (2011). BARD Offshore 1. http://www.bard-offshore.de/en/projects/offshore/bard-offshore-1.html. Accessed March 2014.

Bauman, R. P. (2007). The Bernoulli Conundrum. http://www.introphysics.info. Accessed March 2014.

BBC. (2008). Wind farm's first turbines active. http://news.bbc.co.uk/2/hi/uk_news/england/lincolnshire/7388949.stm. Accessed March 2014.

BBC. (2010). New UK offshore wind farm licences are announced. http://news.bbc.co.uk/2/hi/business/8448203.stm. Accessed March 2014.

BBC. (2011). Region's MPs lobbied over Dogger Bank wind farm. http://www.bbc.co.uk/news/uk-england-15274152. Accessed March 2014.

BBC-News-Technology. (2010). Thanet offshore wind farm starts electricity production. http://www.bbc.co.uk/news/technology-11395824. Accessed March 2014.

bine.info. (2009). RAVE—Research on the offshore test field. http://www.bine.info/en/publications/publikation/rave-forschen-am-offshore-testfeld/?artikel=2214. Accessed March 2014.

blekingeoffshore.se. (2013). Blekinge Offshore. http://www.blekingeoffshore.se/. Accessed March 2014.

Bouhafs, F., & Mackay, M. (2012). Active control and power flow routing in the Smart Grid. IEEE Smart Grid. http://smartgrid.ieee.org/december-2012/734-active-control-and-power-flow-routing-in-the-smart-grid. Accessed March 2014.

Carlin, P. W., Laxson, A. S., & Muljadi, E. B. (2003). The history and state of the art of variable-speed wind turbine technology. *Wind Energy, 6,* 129–159. doi:10.1002/we.77.

c-power.be. (n.d.). Construction of the gravity base foundation. http://www.c-power.be/construction. Accessed March 2014.

c-power.be. (2013). Welcome to C-Power. http://www.c-power.be/English/welcome/algemene_info.html. Accessed March 2014.

Cryan, P. (2013). Wind turbine blade. http://gallery.usgs.gov/photos/10_19_2009_s84Aq11P-Pk_10_19_2009_3. Accessed March 2014.

dantysk.com. (2012). DanTysk Offshore Wind. http://www.dantysk.com/wind-farm/facts-chronology.html. Accessed March 2014.

demowfloat.eu. (n.d.). WindFloat. http://www.demowfloat.eu/. Accessed March 2014.

dnv.com. (2011). New design practices for offshore wind turbine structures. http://www.dnv.com/press_area/press_releases/2011/new_design_practices_offshore_wind_turbine_structures.asp. Accessed March 2014.

dongenergy.com. (n.d.). About Gunfleet Sands. http://www.dongenergy.com/Gunfleetsands/GunfleetSands/AboutGFS/Pages/default.aspx. Accessed March 2014.

dongenergy.com. (2009). About Horns Rev 2. http://www.dongenergy.com/hornsrev2/en/about_horns_rev_2/about_the_project/pages/turbines.aspx. Accessed March 2014.

eastangliawind.com. (2009–2013). East Anglia Offshore Windfarm Zone. http://www.eastangliawind.com/. Accessed March 2014.

ecoseed.org. (2011). South Korea to build $9 billion offshore wind farm by 2019. http://www.ecoseed.org/business/asia/article/130-asia/11823-south-korea-to-build-$9-billion-offshore-wind-farm-by-2019. Accessed March 2014.

Enerpac. (2009). Perpendicular at open sea. http://www.ynfpublishers.com/2009/09/perpendicular-at-open-sea/.

eon.com. (2013). E.ON lays groundwork for Amrumbank West. http://www.eon.com/en/media/news/press-releases/2013/5/29/eon-lays-groundwork-for-amrumbank-west.html. Accessed March 2014.

European-Commission-(FP6). (2011). UpWind, design limits and solutions for very large wind turbines. The sixth framework programme for research and development of the European Commission (FP6).

fukushima-forward.jp. (n.d.). Fukushima FORWARD. http://www.fukushima-forward.jp/. Accessed March 2014.

fukushima-forward.jp. (2013). Semisubmersible. http://www.fukushima-forward.jp/photo/index.html. Accessed March 2014.

KarleHorn. (2010). Windkraftanlage Rotorblatt Achse. de.wikipedia.

London-array-website. (2011). First foundation installed at London Array. http://www.londonarray.com/. Accessed March 2014.

Lorc. (2011b). Knowledge. http://www.lorc.dk/. Accessed March 2014.

lorc.dk. (2011). roedsand-2. http://www.lorc.dk/offshore-wind-farms-map/roedsand-2. Accessed March 2014.

Lorc-website. (2011). http://www.lorc.dk/. Accessed March 2014.

Maciel, J. G. (2010). The WindFloat Project. http://www.wavec.org/content/files/11_Joao_Maciel.pdf. Accessed March 2014.

mhi-global.com. (2013). MHI news. http://www.mhi-global.com/discover/graph/news/no171. html. Accessed March 2014.

mobiusinstitute.com. (n.d.). Vibration analysis of wind turbines. http://www.mobiusinstitute.com/articles.aspx?id=873. Accessed March 2014.

Muljadi, E., & Butterfield, C. P. (1999). Pitch-controlled variable-speed. Presented at the 1999 IEEE Industry Applications, Phoenix, Arizona. NREL/CP-500-27143.

ndr.de. (2013). Rösler eröffnet Offshore-Windpark Bard 1. http://www.ndr.de/regional/niedersachsen/oldenburg/offshore389.html. Accessed March 2014.

nexans.com. (n.d.). Hywind—Floating windpower. http://www.nexans.com/eservice/Corporate-en/navigate_252193/Hywind_floating_windpower.html. Accessed March 2014.

npd.no. (2012). Trollet som ble temmet. http://www.npd.no/Templates/OD/Article.aspx?id=4438. Accessed March 2014.

openei.org. (2010). Vestas. http://en.openei.org/wiki/. Accessed March 2014.

offshorewind.biz. (2012a). UK: Greater Gabbard offshore wind farm generates power. http://www.offshorewind.biz/2012/09/07/uk-greater-gabbard-offshore-wind-farm-generates-power/. Accessed March 2014.

offshorewind.biz. (2012b). UK: Walney Offshore wind farm fully operational. http://www.offshorewind.biz/2012/06/14/uk-walney-offshore-wind-farm-fully-operational/. Accessed March 2014.

offshorewind.biz. (2013a). Construction starts on EnBW Baltic 2 OWF (Germany). http://www.offshorewind.biz/2013/08/20/construction-starts-on-enbw-baltic-2-owf-germany/. Accessed March 2014.

offshorewind.biz. (2013b). Japan floats away from nuclear. http://www.offshorewind.biz/2013/10/31/japan-floats-away-from-nuclear/. Accessed March 2014.

principlepowerinc.com. (n.d.). principlepower. http://www.principlepowerinc.com/. Accessed March 2014.

renewableenergyfocus.com. (2012). Construction starts on Germany's € 1.3bn Meerwind offshore wind farm project. http://www.renewableenergyfocus.com/view/27468/construction-starts-on-germany-s-1-3bn-meerwind-offshore-wind-farm-project/. Accessed March 2014.

renewablesinternational.net. (2013). First floating offshore wind substation underway. http://www.renewablesinternational.net/first-floating-offshore-wind-substation-underway/150/435/71904/. Accessed March 2014.

rt.com. (2013). Japan to start building world's biggest offshore wind farm this summer. http://rt.com/news/japan-renewable-energy-resource-290/. Accessed March 2014.

Ruviaro, M., Rüncos, F., Sadowski, N., & Borges, I. M. (2012). Analysis and test results of a brushless doubly fed induction machine with rotary transformer. *IEEE Transactions on Industrial Electronics, 59*(6), 2670–2677

rwe.com. (2013). Offshore wind farm Nordsee Ost. http://www.rwe.com/web/cms/en/961656/offshore-wind-farm-nordsee-ost/. Accessed March 2014.

rwe.com/. (2013). Gwynt y Môr website. http://www.rwe.com/web/cms/en/1202906/rwe-innogy/sites/wind-offshore/under-construction/gwynt-y-mr/. Accessed March 2014.

siemens.com. (2009a). Hywind: Siemens and StatoilHydro install first floating wind turbine. http://www.siemens.com/press/en/pressrelease/?press=/en/pressrelease/2009/renewable_energy/ere200906064.htm. Accessed March 2014.

Schlipf, D. (2012). Lidar Assisted Control of Wind Turbines. http://rasei.colorado.edu/wind-research-internal1/David_Schlipf.pdf. Accessed March 2014.

seas-nve.dk. (n.d.). Vindeby Offshore Wind Farm. http://www.seas-nve.dk/AboutSeasNve/Wind/References/Offshore/Vindeby.aspx. Accessed March 2014.

smartmeters.com. (2013). Lincs wind farm now fully operational. http://www.smartmeters.com/the-news/renewable-energy-news/4169-lincs-wind-farm-now-fully-operational.html. Accessed March 2014.

stateofgreen.com. (2013). DONG Energy reaches milestone at Borkum Riffgrund 1. http://www.stateofgreen.com/en/Newsroom/DONG-Energy-Reaches-Milestone-at-Borkum-Riffgrund-1. Accessed March 2014.

Statoil-Statkraft. (n.d.). Sheringham Shoal Website. http://www.scira.co.uk/. Accessed March 2014.

trianel-borkum.de. (2013). Windpark. http://www.trianel-borkum.de/de/windpark/daten-und-fakten.html. Accessed March 2014.

VDN. (2007). *TransmissionCode* 2007. Berlin: Verband der Netzbetreiber—VDN—e.V. beim VDEW.

windpowermonthly.com. (2012). Close up—Envision's 3. 6 MW offshore machine. http://www.windpowermonthly.com/article/1115270/close--envisions-36mw-offshore-machine. Accessed March 2014.

windreich.ag. (2013). Global Tech I—Construction Progress. http://windreich.ag/en/global-tech-i-construction-progress/. Accessed March 2014.

Chapter 5

lorc.dk. (2011b). Wave devices. http://www.lorc.dk/wave-energy/list-of-devices.

oceanenergy.ie. (2013). OE buoy. http://www.oceanenergy.ie/about-us/.

Chapter 6

Mocia, J., Arapogianni, A., Wilkes, J., Kjaer, C., & Gruet, R. (2011). *Pure power. Wind energy targets for 2020 and 2030*. Brussels: European Wind Energy Association.

power-technology.com. (n.d.). Green ocean energy wave treader, United Kingdom. http://www.power-technology.com/projects/greenoceanenergywav/greenoceanenergywav1.html.

Chapter 10

Burton, T., Sharpe, D., Jenkins, N., & Bossanyi, E. (2008). *Wind energy handbook*. England: Wiley.

DNV. (2011). *Modelling and analysis of marine operations*. Norway: DNV-RP-H103.

DNV. (2013). *Design of offshore wind turbine structures*. Norway: Det Norske Veritas AS. (January 2013).

Faltinsen, O. (1993). *Sea loads on ships and offshore structures*. UK: Cambridge University Press.

Faltinsen, O. M. (1999). Ringing loads on a slender vertical cylinder of general cross- Section. *Journal of Engineering Mathematics, 35*, 199-217.

Jiang Z. Y., Karimirad M., Moan T. (2012). Response analysis of a parked spar-type wind turbine under different environmental conditions and blade pitch mechanism fault. Greece: ISOPE.

Jonkman, J. M. (2007). *Dynamics modeling and loads analysis of an offshore floating wind turbine*. USA: NREL (TP-500-41958).

Karimirad, M. (2012). Mechanical-dynamic loads. In A. Sayigh (Ed.), *Comprehensive renewable energy* (pp. 243–268). UK: Elsevier.

Karimirad, M. (2013). Modeling aspects of a floating wind turbine for coupled wave-wind-induced dynamic analyses. *Renewable Energy, 53*, 299–305.

Karimirad, M., & Moan, T. (2012a). Stochastic dynamic response analysis of a tension leg spar-type offshore wind turbine. *Journal of Wind Energy (Wiley) 16*(6), 953 -973. September 2013. doi:10.1002/we.

Karimirad, M., & Moan, T. (2012b). A simplified method for coupled analysis of floating offshore wind turbines. *Journal of Marine Structures, 27*, 45–63. doi:10.1016/j.marstruc.2012.03.003.

Karimirad, M., & Moan, T. (2012d). Wave- and wind-induced dynamic response of a spar-type offshore wind turbine. *Journal of Waterway, Port, Coastal, and Ocean Engineering, 138*(1), 9–20. (ASCE).

Karimirad, M., Meissonnier, Q., Gao, Z., & Moan, T. (2011). Hydroelastic code-to-code comparison for a tension leg spar-type floating wind turbine. *Marine Structures,* 412–435.

Korotkin, A. I. (2009). *Added masses of ship structures.* Germany: Springer.

Manwell, J. F., McGowan, J. G., & Rogers, A. L. (2006). *Wind energy explained, theory, design and application.* Chichester: Wiley.

Maruo, H. (1960). The drift of a body floating on waves. *Journal of Ship Research, 4,* 1–10.

Mayilvahanan, A. C., Tørum, A., & Myrhaug, D. (2012). An overview of wave impact forces on offshore wind turbine substructures. *Energy Procedia, 20,* 217–226.

Newman, J. N. (1977). *Marine hydrodynamics.* Cambridge: MIT Press.

Roshko, A. (1960). Experiments on the flow past a circular cylinder at very high Reynolds numbers. *Fluid Mechanics, 10,* 345–356.

Tangler, J. L., & Somers, D. M. (1995). *NREL airfoil families for HAWTs.* Washington D.C.: AWEA.

Chapter 11

blekingeoffshore.se. (2013). *Blekinge Offshore.* http://www.blekingeoffshore.se/. Accessed May 2014.

Jiang, Z. Y., Karimirad, M., & Moan, T. (2012). *Response analysis of a parked spar-type wind turbine under different environmental conditions and blade pitch mechanism fault. ISOPE.* Greece: ISOPE.

Karimirad, M. (2012). Mechanical-Dynamic Loads. In A. Sayigh (Ed.), *Comprehensive renewable energy* (pp. 243–268). Oxford: Elsevier.

Karimirad, M. (2013). Modeling aspects of a floating wind turbine for coupled wave-wind-induced dynamic analyses. *Renewable Energy, 53,* 299–305.

Karimirad, M., & Moan, T. (2010). Effect of aerodynamic and hydrodynamic damping on dynamic response of spar type floating wind turbine. *Proceedings of the EWEC2010, European wind energy conference.* Warsaw: EWEA.

Karimirad, M., & Moan, T. (2012). Stochastic dynamic response analysis of a tension leg spar-type offshore wind turbine. *Journal of Wind Energy* (Wiley). doi: 10.1002/we.

Karimirad, M., Meissonnier, Q., Gao, Z., & Moan, T. (2011). Hydroelastic code-to-code comparison for a tension leg spar-type floating wind turbine. *Marine Structures, 24*(4), 412–435.

Larsen, C. M. (1990). *Response modelling of marine risers and pipelines.* Trondheim: NTNU.

Muliawan, M. J., Karimirad, M., & Moan, T. (2013a). Dynamic response and power performance of a combined Spar-type floating wind turbine and coaxial floating wave energy converter. *Renewable Energy, 50,* 47–57.

Muliawan, M. J., Karimirad, M., Gao, Z., & Moan, T. (2013b). Extreme responses of a combined spar-type floating wind turbine and floating wave energy converter (STC) system with survival modes. *Ocean Engineering, 65,* 71–82.

Woebbeking, M. (2007). *Development of a new standard and innovations in certification of wind turbines.* Germany: GL.

Index

© Springer International Publishing Switzerland 2014
M. Karimirad, *Offshore Energy Structures*, DOI 10.1007/978-3-319-12175-8

Lightning Source UK Ltd.
Milton Keynes UK
UKOW06n0439090616

275915UK00002B/4/P